Everyday Mathematics®

The University of Chicago School Mathematics Project

Student Math Journal
Volume 1

Grade

W0009899

McGraw Hill **Wright Group**

The McGraw-Hill Companies

The University of Chicago School Mathematics Project (UCSMP)

Max Bell, Director, UCSMP Elementary Materials Component; Director, *Everyday Mathematics* First Edition
James McBride, Director, *Everyday Mathematics* Second Edition
Andy Isaacs, Director, *Everyday Mathematics* Third Edition
Amy Dillard, Associate Director, *Everyday Mathematics* Third Edition

Authors

Max Bell
John Bretzlauf
Amy Dillard
Robert Hartfield

Andy Isaacs
James McBride
Kathleen Pitvorec
Peter Saecker

Robert Balfanz*
William Carroll*
Sheila Sconiers*

*First Edition only

Technical Art
Diana Barrie

Photo Credits
© Gregory Adams/Getty Images, cover, *top right;* Getty Images, cover, *center;* ©Tony Hamblin; Frank Lane Picture Agency/CORBIS, cover, *bottom left;* ©Nick Rowe/Getty Images, p. v.

Teachers in Residence
Carla L. LaRochelle, Rebecca W. Maxcy

Editorial Assistant
Laurie K. Thrasher

Contributors
Martha Ayala, Virginia J. Bates, Randee Blair, Donna R. Clay, Vanessa Day, Jean Faszholz, James Flanders, Patti Haney, Margaret Phillips Holm, Nancy Kay Hubert, Sybil Johnson, Judith Kiehm, Deborah Arron Leslie, Laura Ann Luczak, Mary O'Boyle, William D. Pattison, Beverly Pilchman, Denise Porter, Judith Ann Robb, Mary Seymour, Laura A. Sunseri-Driscoll

This material is based upon work supported by the National Science Foundation under Grant No. ESI-9252984. Any opinions, findings, conclusions, or recommendations expressed in this material are those of the authors and do not necessarily reflect the views of the National Science Foundation.

www.WrightGroup.com

Send all inquiries to:
Wright Group/McGraw-Hill
P.O. Box 812960
Chicago, IL 60681

ISBN 0-07-604582-X

13 QWD 12 11 10 09

The **McGraw-Hill** *Companies*

Contents

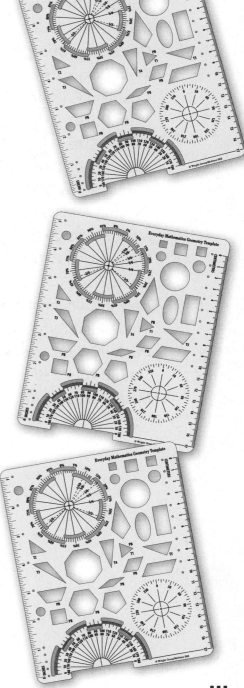

UNIT 2 | Using Numbers and Organizing Data

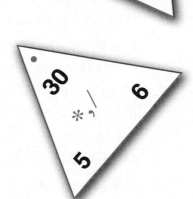

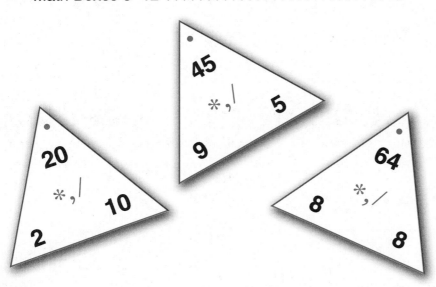

UNIT 4   Decimals and Their Uses

UNIT 5   Big Numbers, Estimation, and Computation

UNIT 6 **Division; Map Reference Frames; Measures of Angles**

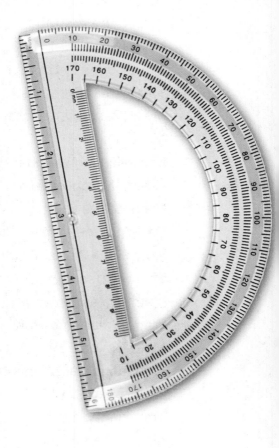

Activity Sheets

Welcome to *Fourth Grade Everyday Mathematics*®

LESSON 1·1

Much of your work in Kindergarten through third grade was basic training in mathematics and its uses. You learned to solve number stories and use arithmetic, including basic addition and multiplication facts.

Fourth Grade Everyday Mathematics builds on this basic training and begins to make the transition to mathematics concepts and ways of using mathematics that are more like what your parents and older siblings may have learned in high school. The authors believe that fourth graders can do more than was thought possible in years past.

Here are some things you will be asked to do in *Fourth Grade Everyday Mathematics:*

◆ Increase your "number sense," "measure sense," and estimation skills.

◆ Extend your skills in the basics of arithmetic—addition, subtraction, multiplication, and division. There is not much more to learn about the arithmetic of whole numbers, but over the next couple of years, you will become comfortable using fractions, percents, and decimals.

◆ Learn about variables (letters that stand for numbers) and other introductory topics in algebra.

◆ Develop your geometry concepts and skills with more exact definitions and classification of geometric figures, constructions and transformations of figures, and investigation of areas and volumes of shapes.

◆ Take a World Tour. Along the way, you will consider many kinds of data about various countries and learn how to use coordinate systems to locate places on world globes and maps.

◆ Do many projects involving numerical data.

In fourth grade, you will be asked to do more independent reading and investigation (often working with partners or in groups) rather than being told everything by your teacher.

We hope that you find the activities fun and that you see the beauty in mathematics. Most importantly, we hope you become better and better at using mathematics to solve interesting problems in your life.

LESSON 1·1   Using Your *Student Reference Book*

Use your *Student Reference Book* to complete the following:

1. Look up the word **mode** in the Glossary.

   a. Copy the definition. _____

   _____

   b. On which page in the *Student Reference Book* could you find more information

   about the mode of a set of data? page _____

2. Find the essay "Comparing Numbers and Amounts."

   a. Describe what you did to find the essay.

   _____

   _____

   b. Read the essay and solve the Check Your Understanding problems.

   Problem 1: _____   Problem 2: _____

   Problem 3: _____   Problem 4: _____

   c. Check your answers using the Answer Key.

3. Look up the rules for the game *Name That Number*.

   a. On which page did you find the rules? page _____

   b. How many players are needed for the game? _____ players

4. Go to the World Tour section. Record two interesting facts you find there.

   a. Fact 1: _____

   I found this information on page _____.

   b. Fact 2: _____

   I found this information on page _____.

Date _____   Time _____

LESSON 1·1   **Math Boxes**

1. Add mentally.

   a. 1 + 7 = _____

   b. 4 + 0 = _____

   c. 5 + 5 = _____

   d. 2 + 9 = _____

   e. 8 + 5 = _____

   f. 7 + 9 = _____

2. Fill in the missing numbers and state the rule.

   a. 2, 4, 6, __8__, _____, _____

   Rule: __+2__

   b. 65, 60, 55, _____, _____, _____

   Rule: _____

   c. 109, 95, 81, _____, _____, _____

   Rule: _____

3. Complete.

21 in. = _____ ft _____ in.

Circle the best answer.

   A. 1 ft 1 in.

   B. 1 ft 10 in.

   C. 1 ft 9 in.

   D. 1 ft 3 in.

4. Complete.

   a. 2 quarters = _____ dimes

   b. 1 dollar and 5 nickels = _____ quarters

   c. 14 dimes = _____ pennies

   d. 8 quarters = _____ dollars

   e. 3 quarters and 9 nickels = _____ dimes

5. Add mentally or with a paper-and-pencil algorithm.

   a. 32 + 35 = _____

   b. 38 + 66 = _____

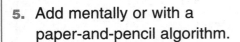
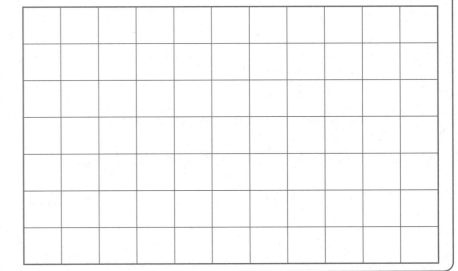

LESSON 1·2 — Points, Line Segments, Lines, and Rays

Use a straightedge to draw the following:

SRB
90 91

1. a. Draw and label line segment RT ($\overline{RT}$).

   b. What is another name for $\overline{RT}$? _____

2. a. Draw and label line BN ($\overleftrightarrow{BN}$). Draw and label point T on it.

   b. What are 2 other names for $\overleftrightarrow{BN}$? _____

3. a. Draw and label ray SL ($\overrightarrow{SL}$). Draw and label point R on it.

   b. What is another name for $\overrightarrow{SL}$? _____

4. a. Draw a line segment from each point to each of the other points.

   M•   N•

   •O   •P

   b. How many line segments did you draw? _____

   c. Write a name for each line segment you drew.

   _____

LESSON 1·2 Math Boxes

1. Subtract mentally.

a. $6 - 3 =$ _____

b. $7 - 4 =$ _____

c. $14 - 7 =$ _____

d. $16 - 9 =$ _____

e. _____ $= 9 - 4$

f. _____ $= 17 - 9$

2. Draw and label line *QR*.
Draw point *S* on it.

What are two other names for line *QR*?

_____

_____

SRB
91

3. Complete.

Max read _____ books.

Sue read _____ books.

Ira read _____ books.

Pat read _____ books.

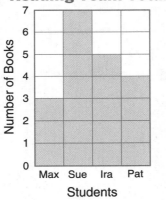

Reading Team Totals

Number of Books (y-axis: 0–7)

Students (x-axis: Max, Sue, Ira, Pat)

SRB
76

4. Cross out the names that do not belong in the name-collection box. Label the box with the correct number.

$25 - 13$
$20 - 7$
6×2
4×3
$40 - 23$
7×3

SRB
149

5. Subtract mentally or with a paper-and-pencil algorithm.

a. $86 - 21 =$ _____

b. $93 - 24 =$ _____

SRB
12–15

LESSON 1·3 **Angles**

1. Which angle is bigger,

   ∠ABC or ∠DEF? _____

2. Draw ∠BAC. What is another name

   for ∠BAC? _____

   C•

3. What is the vertex of ∠BAC? Point _____

   A•   •B

4. Feng said the name of this angle is ∠SRT. Is he right? Explain.

   _____

   _____

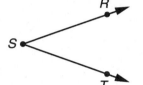

Try This

Use the points shown on the grid below and a straightedge to draw the angles.

5. Draw ∠AED.

   a. What is the vertex of the angle?   Point _____

   b. What is another name for ∠AED?   ∠ _____

6. Draw a right angle whose vertex is point C.   My angle is called ∠ _____.

7. Draw an angle that is smaller than a right angle.   My angle is called ∠ _____.

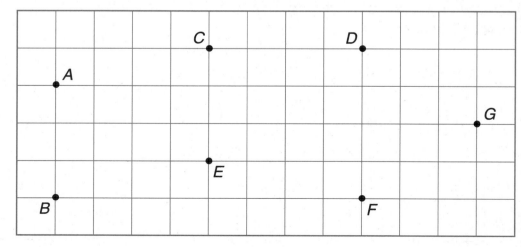

Addition and Subtraction

Use your favorite addition and subtraction methods to solve the problems. Show your work.

SRB
10–15

1. 48 + 79 = _____

2. 392 + 75 = _____

3. _____ = 296 + 514

4. _____ = 937 + 652

5. _____ = 98 − 57

6. 122 − 82 = _____

7. _____ = 900 − 632

8. _____ = 512 − 239

LESSON 1·3   **Math Boxes**

---

1. Add mentally.

   a. 2 + 5 = _____

   b. 3 + 3 = _____

   c. 4 + 6 = _____

   d. _____ = 6 + 7

   e. _____ = 9 + 8

   f. _____ = 4 + 7

2. Fill in the missing numbers and state the rule.

   a. 4, 8, 12, 16, _____, _____, _____

   Rule: _____

   b. 33, 30, 27, _____, _____, _____

   Rule: _____

   c. _____, _____, _____, 106, 141, 176

   Rule: _____

   SRB
   160 161

3. Complete.

   a. 1 ft = _____ in.

   b. 24 in. = _____ ft

   c. _____ yd = 36 in.

   d. 30 in. = _____ ft _____ in.

   e. 50 in. = _____ yd _____ ft

      _____ in.

   SRB
   129

4. Complete.

   a. 9 dimes = _____ pennies

   b. 30 dimes = _____ dollars

   c. 4 quarters = _____ dimes

   d. 2 dollars
      and 10 nickels = _____ quarters

   e. 13 dollars = _____ quarters

5. Add mentally or with a paper-and-pencil algorithm.

   a. 63 + 12 = _____

   b. 56 + 97 = _____

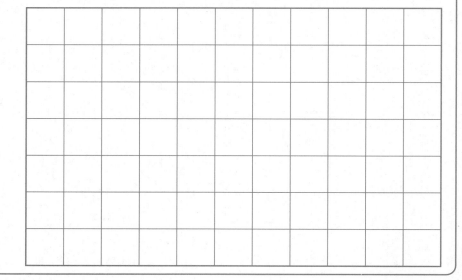

SRB
10 11

LESSON 1·4   **Math Boxes**

1. Subtract mentally.

   a. 10 − 4 = _____

   b. _____ = 8 − 5

   c. 7 − 4 = _____

   d. 15 − 7 = _____

   e. 13 − 8 = _____

   f. _____ = 17 − 8

2. Draw and label line *AB*.
   Draw point *C* on it.

   What are two other names for line *AB*?

   _____

   _____

3. Complete.

   Luz sold _____ boxes.

   Ana sold _____ boxes.

   Mya sold _____ boxes.

   Pei sold _____ boxes.

   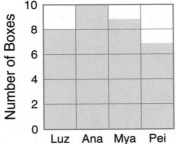

4. Which of these can go in a name-collection box for the number 50? Circle the best answer.

   A. 10 + 35

   B. 136 − 51

   C. 200 ÷ 4

   D. 4 × 15

5. Subtract mentally or with a paper-and-pencil algorithm.

   a. 76 − 41 = _____

   b. 52 − 38 = _____

   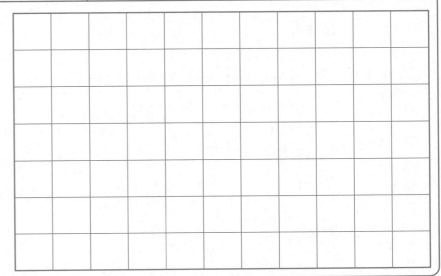

LESSON 1·4   **Parallelograms**

1. Circle the pairs of line segments below that are parallel. Check some of your answers by extending each pair of segments to see if the two segments in the pair meet or cross.

a. _____

b.

c.

d.

e. _____

f.

Use your Geometry Template or straightedge to draw the following quadrangles:

2. Draw a quadrangle that has 2 pairs of parallel sides.

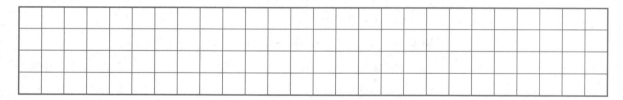

This is called a _____.

3. Draw a quadrangle that has only 1 pair of parallel sides.

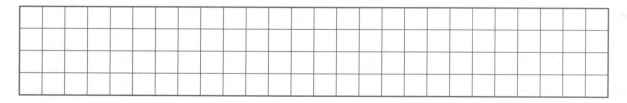

This is called a _____.

LESSON 1·4 Parallelograms *continued*

For Problems 4 and 5, circle the best answer(s). Some items have more than
1 correct answer, so you may need to circle more than 1 answer.

4. A **parallelogram** is a quadrangle that
   has 2 pairs of parallel sides.
   Which are parallelograms?

   A. squares

   B. rectangles

   C. rhombuses

   D. trapezoids

5. A **rhombus** is a parallelogram in which
   all sides are the same length.
   Which are always rhombuses?

   A. squares

   B. rectangles

   C. trapezoids

   D. kites

Try This

A **rectangle** is a parallelogram that has all right angles. Which of the following
are rectangles? Write *always, sometimes,* or *never* to complete each sentence.
Explain your answers.

6. Squares are _____ rectangles. Explain. _____

   _____

7. Rhombuses are _____ rectangles. Explain. _____

   _____

8. Trapezoids are _____ rectangles. Explain. _____

   _____

9. A kite is _____ a parallelogram. Explain. _____

   _____

LESSON 1·5 **What Is a Polygon?**

These are polygons.

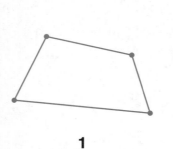
1

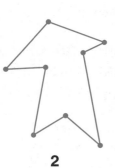
2

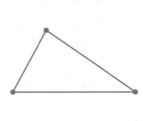
3

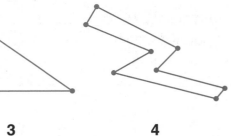
4

These are NOT polygons.

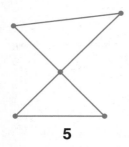
5

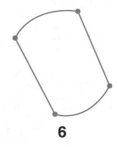
6

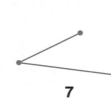
7

8

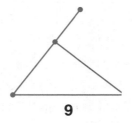
9

1. If you had to explain what a polygon is, what would you say? (*Think:* What do Polygons 1–4 have in common? How are Shapes 5–9 different from Polygons 1–4?)

_____

_____

_____

_____

_____

2. Choose one of the shapes from above. Explain why the shape is not a polygon.

_____

_____

_____

_____

_____

LESSON
1·5

Math Boxes

---

1. Subtract mentally.

   a. $7 - 0 =$ _____

   b. $10 - 7 =$ _____

   c. _____ $= 9 - 4$

   d. _____ $= 14 - 6$

   e. $13 - 7 =$ _____

   f. $16 - 9 =$ _____

---

2. Draw $\angle MRT$.

                                     T
                                       •

    M •

                          •
                          R

What is another name for $\angle MRT$?

_____

---

3. Draw and label line segment *AB*.

What is another name for $\overline{AB}$?

_____

---

4. Name as many rays as you can in the figure below.

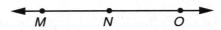

        M   *N*   *O*

Write their names.

_____

_____

_____

---

5. Which polygons have 2 pairs of parallel sides? Circle the best answer.

   A. square and trapezoid

   B. rectangle and rhombus

   C. triangle and parallelogram

   D. pentagon and square

---

6. Put these numbers in order from least to greatest.

    10,005   51,000

    5,100   10,500

    _____

    _____

    _____

    _____

LESSON 1·6 An Inscribed Square

Follow the directions below to make a square that you will tape on the next page.

Step 1 Use your compass to draw a circle on a sheet of colored paper. The circle should be small enough to fit on the next page. Cut out the circle.

Step 2 With your pencil, make a dot in the center of the circle, where the hole is, on both the front and the back.

Step 3 Fold the circle in half. Make sure that the edges match and that the fold line passes through the center. Be sure to make sharp creases.

Step 4 Fold the folded circle in half again so that the edges match.

Step 5 Unfold your circle. The folds should pass through the center of the circle and form 4 right angles.

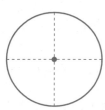

Step 6 Using a straightedge, connect the endpoints of the folds at the edge of the circle to make a square. Cut out the square.

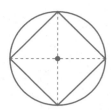

LESSON 1·6 **An Inscribed Square** *continued*

Now use your compass to find out whether the 4 sides of your square are about the same length.

Place the anchor on one endpoint of a side and the pencil point on the other endpoint of the side. Then, *without changing the compass opening,* try to place the anchor and pencil point on the endpoints of each of the other sides.

If the sides of your square are about the same length, tape the square in the space below. If not, follow the directions on page 14 again. Tape your best square in the space below.

LESSON 1·6   # Math Boxes

1. Subtract mentally.

    a. $9 - 5 =$ _____

    b. $11 - 2 =$ _____

    c. _____ $= 14 - 7$

    d. _____ $= 12 - 4$

    e. $13 - 6 =$ _____

    f. _____ $= 12 - 9$

2. Which of the shape(s) below are NOT polygons? _____

    A   B   C

SRB 96

3. Draw a quadrangle with only 1 right angle. Draw in the right angle symbol.

How do you know it is a right angle?

_____

SRB 93 99

4. Circle the convex polygon(s).

SRB 97

5. Draw and label ray *HA*.
Draw point *T* on it.

What is another name for $\overrightarrow{HA}$? _____

SRB 91

6. In the numeral 42,318, the 2 stands for 2,000.

    a. The 1 stands for _____.

    b. The 8 stands for _____.

    c. The 4 stands for _____.

    d. The 3 stands for _____.

SRB 4

LESSON 1·7   Circle Constructions

Do each of the following 3 constructions on a separate sheet of paper. Try and try again until you are satisfied with your work. Then cut out your 3 best constructions and tape them in your journal.

1. Use your compass to draw a picture of a circular dartboard. It is not necessary to include the details of the board. Tape your best work in the space below. The circles in the dartboard and in your picture are called **concentric circles.**

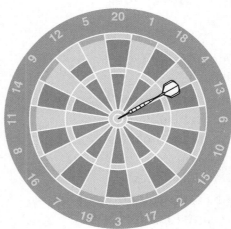

Circle Constructions *continued*

2. a. Make a dot near the center of
your paper. Use your compass
to draw a circle with that dot as
its center.

b. *Without changing the opening
of your compass,* draw a
congruent circle that **intersects**
the center of the first circle. Mark
the center of the second circle.

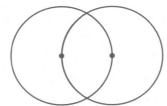

c. *Without changing the opening
of your compass,* draw a third
congruent circle that intersects
the center of each of the first
2 circles.

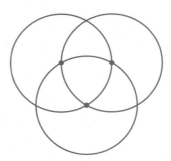

Try and try again until you are satisfied
with your work. Then cut out your circle
design and tape it in the space below.

LESSON 1·7 **Circle Constructions** *continued*

Try This

3. Draw this design with your compass. Work on separate sheets of paper until you are satisfied with your work. Color your best design. Then cut it out and tape it in the space below.

   Hint: Start by making the 3-circle design on page 18. Then add more circles to it.

LESSON 1·7 **Math Boxes**

---

1. Subtract mentally.

a. $9 - 7 =$ _____

b. $10 - 6 =$ _____

c. _____ $= 16 - 8$

d. _____ $= 17 - 7$

e. $13 - 7 =$ _____

f. _____ $= 15 - 9$

---

2. Draw ∠TIF. What is the vertex of ∠TIF?

Point _____

F•

•T

I•

92

---

3. Draw and label line segment GP.

What is another name for $\overline{GP}$?

_____

90

---

4. Name as many rays as you can in the figure below.

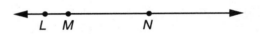
L M   N

Write their names.

_____

_____

91

---

5. Draw a quadrangle with 1 pair of parallel sides.

What kind of quadrangle is this?

_____

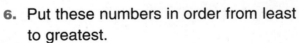
99 100

---

6. Put these numbers in order from least to greatest.

32,000   3,200

23,000   2,300

_____

_____

_____

_____

4

LESSON 1·8

Copying a Line Segment

Steps 1–4 below show you how to copy a line segment.

Step 1 You are given line segment *AB* to copy.

Step 2 Draw a line segment that is longer than line segment *AB*. Label one of its endpoints *C*.

Step 3 Open your compass so that the anchor is on one endpoint of line segment *AB* and the pencil point is on the other endpoint.

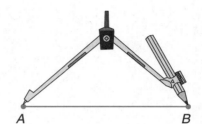

Step 4 *Without changing the compass opening,* place the anchor on point *C* on your second line segment. Make a mark that crosses this line segment. Label the point where the mark crosses the line segment with the letter *D*.

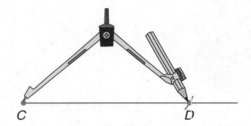

Line segment *CD* should be about the same length as line segment *AB*. Line segments *CD* and *AB* are **congruent.**

Use a compass and straightedge to copy the line segments shown below. For each problem, begin by drawing a line segment that is longer than the one given.

1.

2.   *M*   *N*

Hexagons in Our World

Hexagons are seen in the natural world and in things that people make and use. For example, bees make honeycombs with hexagonal shapes, and snowflakes suggest the shape of a hexagon.

Soccer balls are made up of regular hexagons and regular pentagons.

For many centuries, wonderful tile designs have been created all over the world, especially in Islamic art. As these pictures show, tile designs can be developed in many ways from a pattern that uses hexagons.

Many quilt and fabric designs come from dividing regular hexagons into triangles or rhombuses. You may have made designs like these with pattern blocks. Coloring a design often makes the design more interesting.

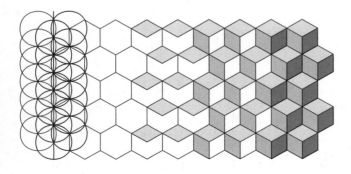

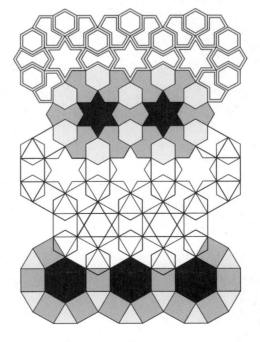

LESSON 1·8

Constructing an Inscribed, Regular Hexagon

Follow each step below. Draw on a separate sheet of paper. Repeat these steps several times. Cut out your best work, and tape it onto the bottom of this page.

Step 1 Draw a circle. (Keep the same compass opening for Steps 2 and 3.) Draw a dot on the circle. Place the anchor of your compass on the dot and make a mark on the circle.

Step 2 Place the anchor of your compass on the mark you just made and make another mark on the circle.

Step 3 Do this four more times to divide the circle into 6 equal parts. The 6th mark should be on the dot you started with or very close to it.

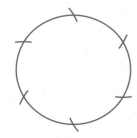

Step 4 With your straightedge, connect the 6 marks on the circle to form a regular hexagon. Use your compass to check that the sides of the hexagon are all about the same length.

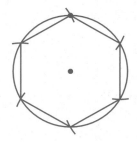

The hexagon is **inscribed** in the circle because each vertex of the hexagon is on the circle.

More Constructions

Construct a regular hexagon on a separate sheet of paper. Then divide the hexagon into 6 equilateral triangles. Use your compass to check that the sides of the 6 equilateral triangles are all about the same length.

Try this several times until you are satisfied with your work. Then cut out your best work and tape it in the space below.

LESSON 1·8 Definition Match

Match each description of a geometric figure in Column I with its name in Column II. Some of the items in Column II do not have a match.

I	II
a. a polygon with 4 right angles and 4 sides of the same length	_____ octagon
	_____ rhombus
b. a polygon with 4 sides, none of which are the same length	_____ right angle
c. a quadrilateral with exactly 1 pair of opposite sides that is parallel	_____ trapezoid
	_____ hexagon
d. lines that never intersect	_____ square
e. a parallelogram with all sides the same length, but not a rectangle	_____ equilateral triangle
f. a polygon with 8 sides	_____ perpendicular lines
g. a polygon with 5 sides	_____ parallel lines
	_____ pentagon
h. an angle that measures 90°	_____ isosceles triangle
i. a triangle with all sides the same length	_____ quadrangle

LESSON 1·8   Math Boxes

1. Subtract mentally.

  a. 14 − 9 = _____

  b. 13 − 8 = _____

  c. _____ = 18 − 9

  d. 17 − 8 = _____

  e. _____ = 11 − 7

  f. _____ = 15 − 6

2. Which of the shape(s) below are

polygons? _____

  A   B   C

SRB
96

3. Draw a quadrangle that has 2 pairs of parallel sides and no right angles.

What kind of quadrangle is this?

_____

SRB
99 100

4. Circle the concave (nonconvex) polygon(s).

SRB
97

5. Draw and label ray *CA*.
Draw point *R* on it.

What is another name for ray *CA*?

_____

SRB
91

6. In the numeral 30,516, what does the 3 stand for? Circle the best answer.

  A. 3,000

  B. 30

  C. 30,000

  D. 300,000

SRB
4

LESSON 1·9 Math Boxes

1. Add.

   a.   64   b.   48
     + 32   + 96

SRB
10 11

2. Subtract.

   a.   78   b.   81
     − 42   − 36

SRB
12–15

3. Put these numbers in order from least to greatest.

46,000   40,600

4,600   4,006

_____

_____

_____

_____

SRB
4

4. In the numeral 78,965,

   a. the 8 stands for _____.

   b. the 6 stands for _____.

   c. the 7 stands for _____.

   d. the 9 stands for _____.

SRB
4

5. Use the following list of numbers to answer the questions:

12, 3, 15, 6, 12, 14, 6, 5, 9, 12

   a. Which number is the least? _____

   b. Which number is the greatest? _____

   c. What is the difference between the least and greatest numbers? _____

   d. Which number appears most often? _____

SRB
73

LESSON 2·1   **A Visit to Washington, D.C.**

Refer to pages 267–269 in your *Student Reference Book*.

SRB
267–269

1. About how many people tour the White House every year?
   Check the best answer.

   _____ between 100 thousand and 1 million   _____ between 1 million and 10 million

   _____ between 10 million and 100 million   _____ between 100 million and 1 billion

2. About how many people ride the Washington Metrorail on an average weekday?
   Check the best answer.

   _____ between 100 thousand and 1 million   _____ between 1 million and 10 million

   _____ between 10 million and 100 million   _____ between 100 million and 1 billion

3. The Library of Congress adds about _____ items each day. About how many

   days does it take to add 50,000 items to the Library of Congress? _____

4. Write the year that each event happened. Then draw a dot for each event
   on the timeline below. Label the dot with the correct letter and date.

   A The year the Metrorail opened ___*1976*___

   B The year of the flight of the *Flyer* _____

   C The year the Washington Monument was completed _____

   D The year the Lincoln Memorial was dedicated _____

   E The year of the first nonstop flight across the Atlantic _____

   F The year the Jefferson Memorial was dedicated _____

   G The year of the first landing on the moon _____

LESSON 2·1   **Math Boxes**

---

1. Add mentally.

   a. 3 + 5 = _____

   b. 30 + 50 = _____

   c. 300 + 500 = _____

   d. _____ = 9 + 7

   e. _____ = 90 + 70

   f. _____ = 900 + 700

SRB 10 11

---

2. What is the value of the digit 5 in 5**60**? ____500____

What is the value of the digit 7 in the numbers below?

   a. 474 _____

   b. 70,158 _____

   c. 187,943 _____

   d. 2,731,008 _____

SRB 4

---

3. Circle the pair of concentric circles.

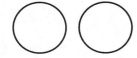

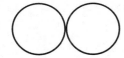

---

4. I am a 2-dimensional figure.
I have two pairs of parallel sides.
All my angles have the same measure.
All my sides are the same length.

What am I? _____

Use your Geometry Template to draw me.

SRB 100

---

5. A giant tortoise can live for about 150 years. An elephant can live for about 78 years. About how much longer can a giant tortoise live than an elephant? Fill in the circle next to the best answer.

   Ⓐ 88 years

   Ⓑ 238 years

   Ⓒ 72 years

   Ⓓ 228 years

SRB 178

---

6. Multiply mentally.

   a. 2 × 1 = _____

   b. _____ = 5 × 0

   c. _____ = 5 × 2

   d. 5 × 4 = _____

   e. 3 × 10 = _____

SRB 16

LESSON 2·2   Name-Collection Boxes

Write five names in each box below. Use as many different kinds of numbers (such as counting numbers, fractions, decimals, negative numbers) and different operations (+, −, ×, ÷) as you can. Draw a star next to the name you find most interesting.

SRB
149

1.

16
8 × 2
970 − 954
10 + (60 ÷ 10)

2.

24

3.

50

4.

100

Make up your own name-collection boxes. Use different kinds of numbers and operations.

5.

6.

LESSON 2·2 Math Boxes

---

1. a. Write the largest number you can make with the digits 5, 2, 3, 0, 6, 0. Use each digit only once.

_____

b. Use the same digits and write the smallest number you can make. Do not start with 0.

_____

SRB
4

---

2. Add mentally or with a paper-and-pencil algorithm.

a.   37
  + 142

b.   468
  + 394

SRB
10 11

---

3. Draw a convex polygon.

SRB
97

---

4. Measure these line segments to the nearest centimeter.

a. _____

    About _____ centimeters

b. _____

    About _____ centimeters

SRB
128

---

5. Complete.

a. 4 ft = _____ in.

b. 4 ft = _____ yd _____ in.

c. 2 yd = _____ ft

d. 72 in. = _____ yd _____ ft

e. 6,756 in. = _____ ft

SRB
129

---

6. Divide mentally.

a. $9 \div 9 =$ _____

b. _____ $= 12 \div 2$

c. _____ $= 20 \div 5$

d. $40 \div 10 =$ _____

e. $14 \div 7 =$ _____

SRB
20

LESSON 2·3

Place-Value Chart

SRB 4

Number											
Hundred Millions 100M											
Ten Millions 10M											
Millions M											
Hundred Thousands 100K											
Ten Thousands 10K											
Thousands K											
Hundreds H											
Tens T											
Ones O											

2·3 Taking Apart, Putting Together

Complete.

1. In 574

5 is worth _____ 500 _____

7 is worth _____

4 is worth _____

2. In 9,027

9 is worth _____

0 is worth _____

2 is worth _____

3. In 280,743

8 is worth _____

2 is worth _____

4 is worth _____

4. In 56,010,837

6 is worth _____

1 is worth _____

5 is worth _____

5. In 705,622,463

5 is worth _____

6 is worth _____

7 is worth _____

6. In 123,456,789

4 is worth _____

3 is worth _____

2 is worth _____

Add.

7.
```
    900
     70
+     5
```

8.
```
  30,000
   7,000
      50
+      2
```

9.
```
  50,000,000
   9,000,000
      60,000
       2,000
         800
+         50
```

10.
```
  300,000,000
    9,000,000
      200,000
       70,000
           30
+           1
```

11. Think about why we need zeros when writing numbers. What would happen if you did not write the zero in the number 5,074?

_____

_____

Date _____ Time _____

 **Polygon Checklist**

Place a check mark next to all of the statements that are true about each figure.
Write an additional true statement for each figure.

1.

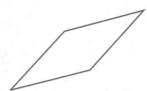

_____ 1 pair of parallel sides

_____ at least 1 right angle

_____ quadrangle

_____ polygon

_____ concave

_____ parallelogram

✓ _____

2.

_____ 4 sides of equal length

_____ kite

_____ square

_____ parallelogram

_____ convex

_____ opposite sides parallel

✓ _____

3.

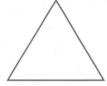

_____ all sides of equal length

_____ all angles of equal measure

_____ one right angle

_____ polygon

_____ equilateral triangle

_____ 1 pair of parallel sides

✓ _____

4.

_____ regular polygon

_____ all sides of equal length

_____ all angles of equal measure

_____ pentagon

_____ octagon

_____ all angles smaller than right angles

✓ _____

LESSON 2·3 Math Boxes

1. Add mentally.

  a. 4 + 5 = _____

  b. 40 + 50 = _____

  c. 400 + 500 = _____

  d. _____ = 5 + 8

  e. _____ = 50 + 80

  f. _____ = 500 + 800

SRB
10 11

2. What is the value of the digit 8 in the numbers below?

  a. 5**8**4 _____

  b. 3**8**,067 _____

  c. 49,**8**41 _____

  d. **8**20,731 _____

  e. **8**,391,467 _____

SRB
4

3. Use your compass to draw a pair of concentric circles.

4. I am a 2-dimensional figure.
I have two pairs of parallel sides.
None of my angles is a right angle.
All of my sides are the same length.

What am I? _____

Use your Geometry Template to draw me.

SRB
100

5. A sailfish can swim at a speed of 110 kilometers per hour. A tiger shark can swim at a speed of 53 kilometers per hour. How much faster can a sailfish swim than a tiger shark?

_____ kilometers per hour

6. Multiply mentally.

  a. 8 × 1 = _____

  b. _____ = 9 × 0

  c. _____ = 5 × 6

  d. 5 × 5 = _____

  e. 7 × 10 = _____

SRB
16

LESSON 2·4 Calculator "Change" Problems

1. Follow your teacher's directions to complete the "change" problems below.
   Use your calculator.

SRB
4

	Start with	Place of Digit	Change to	Operation	New Number
a.	570	Tens			
b.	409	Hundreds			
c.	54,463	Thousands			
d.	760,837	Tens			
e.	52,036,458	Ones			
f.		Ten Thousands			
g.		Millions			

2. Complete these calculator "change" problems on your own.

	Start with	Place of Digit	Change to	Operation	New Number
a.	893	Tens	3	−	
b.	5,489	Hundreds	7	+	
c.	94,732	Thousands	6	+	
d.	218,149	Ten Thousands	0	−	
e.	65,307,000	Millions	9	+	
f.	873,562,003	Ten Millions	1		
g.	103,070,651	Hundred Millions	8		

LESSON 2·4 — Math Boxes

1. What is the largest number you can make with the digits 3, 0, 3, 8, and 0? Fill in the circle next to the best answer.

Ⓐ 83,003

Ⓑ 83,030

Ⓒ 83,300

Ⓓ 80,033

2. Add mentally or with a paper-and-pencil algorithm.

a.
```
   145
 +  34
```

b.
```
   297
 + 136
```

3. Draw a concave pentagon.

4. Measure these line segments to the nearest centimeter.

a.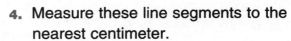

About _____ centimeters

b. _____

About _____ centimeters

5. Complete.

a. 14 in. = _____ ft _____ in.

b. _____ in. = 2 ft

c. _____ ft = 7 yd

d. 1 yd 1 ft = _____ in.

e. 413 ft = _____ yd _____ ft

6. Divide mentally.

a. 16 ÷ 2 = _____

b. 20 ÷ 10 = _____

c. _____ = 40 ÷ 5

d. 60 ÷ 10 = _____

e. _____ = 45 ÷ 5

LESSON 2·5 Counting Raisins

1. Use your $\frac{1}{2}$-ounce box of raisins. Complete each step when the teacher tells you. Stop after you complete each step.

   a. Don't open your box yet. **Guess** about how many raisins are in the box.

      About _____ raisins

   b. Open the box. Count the number of raisins in the top layer. Then **estimate** the total number of raisins in the box.

      About _____ raisins

   c. Now **count** the raisins in the box.

      How many? _____ raisins

2. Make a tally chart of the class data.

Number of Raisins	Number of Boxes

3. Find the following **landmarks** for the class data.

   a. What is the **maximum,** or largest, number of raisins found? _____

   b. What is the **minimum,** or smallest, number of raisins found? _____

   c. What is the **range?** (Subtract the minimum from the maximum.) _____

   d. What is the **mode,** or most frequent number of raisins found? _____

Try This

4. What is the **median** number of raisins found? _____

5. What is the **mean** number of raisins found? _____

Math Boxes

1. A number has

6 in the hundreds place,
1 in the millions place,
2 in the tens place,
8 in the hundred-thousands place,
5 in the ones place,
3 in the thousands place, and
4 in the ten-thousands place.

Write the number.

___, ___ ___ ___, ___ ___ ___

SRB
4

2. Write five names for 34.

34

SRB
149

3. Write >, <, or = to make each number sentence true.

a. 14 _____ 26

b. 3,003 _____ 3,300

c. 12 + 12 _____ 24

d. 200 − 50 _____ 100

e. 30 + 30 _____ 50 + 10

SRB
148 149

4. Name the two pairs of parallel sides in parallelogram *HIJK*.

_____ and _____

_____ and _____

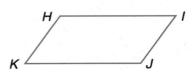

SRB
94

5. Measure these line segments to the nearest $\frac{1}{2}$ centimeter.

a.

About _____ centimeters

b. _____

About _____ centimeters

SRB
128

6. Multiply mentally.

a. 5 × 7 = _____

b. 3 × _____ = 18

c. _____ × 7 = 56

d. 9 × _____ = 45

e. 8 × 4 = _____

SRB
16

LESSON
2·6   **Family Size**

Follow your teacher's directions and complete each step.

1. How many people are in your family? _____ people
   Write the number on a stick-on note.

2. Make a line plot of the family-size data for the class.
   Use **X**s in place of stick-on notes.

Class Data on Family Size

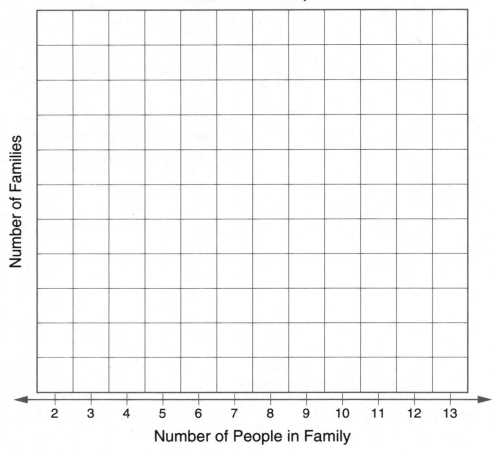

Number of Families

2   3   4   5   6   7   8   9   10   11   12   13

Number of People in Family

3. Find the following landmarks for the class data:

   a. What is the **maximum** (largest) number of people in a family? _____ people

   b. What is the **minimum** (smallest) number of people in a family? _____ people

   c. What is the **range?** (Subtract the minimum from the maximum.) _____ people

   d. What is the **mode** (most frequent family size)? _____ people

4. What is the **median** family size for the class? _____ people

LESSON 2·6 Math Boxes

1. Add mentally.

a. 2 + 7 = _____

b. 20 + 70 = _____

c. 200 + 700 = _____

d. _____ = 8 + 4

e. _____ = 80 + 40

f. _____ = 800 + 400

SRB
10 11

2. Find the median of the data set.

3, 45, 13, 15, 3, 7, 19

Fill in the circle next to the best answer.

Ⓐ 3

Ⓑ 13

Ⓒ 15

Ⓓ 45

SRB
73

3. Subtract mentally or with a paper-and-pencil algorithm.

a.   147
    − 56
   _____

b.   531
    − 246
   _____

SRB
12–15

4. Write 4,007,392 in words.

_____

_____

_____

_____

_____

SRB
4

5. A royal python can be 35 feet long. An anaconda can be 28 feet long. What would be their combined length, end-to-end?

_____ feet

6. Tell whether each number sentence is true or false.

a. 14 + 7 = 22   _____

b. 36 = 15 + 5   _____

c. 45 − 12 = 33   _____

d. 27 = 40 − 13   _____

SRB
148

LESSON
2·7
Partial-Sums Addition

Write a number model for a ballpark estimate. Solve Problems 1–3 using the partial-sums method. Solve Problems 4–6 using any method. Compare your answer with your estimate to see if your answer makes sense.

1.
 	
    76
 + 38
	2.
	
	   647
	 + 936
		3.
		
		  1,672
		 + 3,221

Ballpark estimate: _____	Ballpark estimate: _____	Ballpark estimate: _____

4.
 	
   66
 + 28
	5.
	
	  736
	 + 645
		6.
		
		  7,854
		 + 4,550

Ballpark estimate: _____	Ballpark estimate: _____	Ballpark estimate: _____

Try This

7. Name three 4-digit numbers whose sum is 17,491.

_____ + _____ + _____ = 17,491

LESSON 2·7 Column Addition

Write a number model for a ballpark estimate. Solve Problems 1–3 using the column-addition method. Solve Problems 4–6 using any method. Compare your answer with your estimate to see if your answer makes sense.

1. $\begin{array}{r} 94 \\ + 47 \\ \hline \end{array}$ Ballpark estimate: _____	**2.** $\begin{array}{r} 385 \\ + 726 \\ \hline \end{array}$ Ballpark estimate: _____	**3.** $\begin{array}{r} 2,538 \\ + 4,179 \\ \hline \end{array}$ Ballpark estimate: _____
4. $\begin{array}{r} 49 \\ + 33 \\ \hline \end{array}$ Ballpark estimate: _____	**5.** $\begin{array}{r} 469 \\ + 946 \\ \hline \end{array}$ Ballpark estimate: _____	**6.** $\begin{array}{r} 4,614 \\ + 6,058 \\ \hline \end{array}$ Ballpark estimate: _____

Try This

7. Name four 4-digit numbers whose sum is 15,706.

_____ + _____ + _____ + _____ = 15,706

LESSON 2·7

Math Boxes

---

1. A number has

3 in the millions place,
1 in the ones place,
8 in the thousands place,
9 in the ten-thousands place,
0 in the tens place,
6 in the hundred-thousands place, and
5 in the hundreds place.

Write the number.

___, ___ ___ ___, ___ ___ ___

SRB
4

---

2. Write five names for 100.

100

SRB
149

---

3. Write >, <, or = to make each number sentence true.

a. 16 + 11 _____ 47

b. 206 _____ 602

c. 150 − 50 _____ 100

d. 62 + 10 + 10 _____ 62 − 10 − 10

e. 423,726 _____ 413,999

SRB
148 149

---

4. Draw a parallelogram. Label the vertices so that side *AB* is parallel to side *CD*.

SRB
99 100

---

5. Measure these line segments to the nearest $\frac{1}{2}$ centimeter.

a. _____

About _____ cm

b. _____

About _____ cm

SRB
128

---

6. Multiply mentally.

a. 5 × 8 = _____

b. 2 × _____ = 16

c. 7 × _____ = 21

d. _____ × 9 = 54

e. 8 × 3 = _____

SRB
16

LESSON 2·8 Math Boxes

1. Add mentally.

a. 2 + 4 = _____

b. 20 + 40 = _____

c. 200 + 400 = _____

d. _____ = 8 + 6

e. _____ = 80 + 60

f. _____ = 800 + 600

SRB
10 11

2. Find the following landmarks for this set of numbers: 12, 16, 23, 15, 16, 19, 18.

a. median _____

b. mode _____

c. maximum _____

d. minimum _____

e. range _____

f. mean _____

SRB
73–75

3. Subtract mentally or with a paper-and-pencil algorithm.

a.   231
  –   84

b.   603
  – 466

SRB
12–15

4. Write 8,042,176 in words.

_____

_____

_____

_____

_____

SRB
4

5. An ostrich can weigh about 345 pounds. An emu can weigh about 88 pounds. How much would they weigh together?

_____ pounds

6. Tell whether each number sentence is true or false.

a. 18 + 9 = 37 _____

b. 29 = 17 + 12 _____

c. 42 – 15 = 27 _____

d. 17 = 40 – 24 _____

e. 154 – 65 = 99 _____

SRB
148

LESSON 2·8 Head Sizes

Ms. Woods owns a clothing store. She is trying to decide how many children's hats to stock in each possible size. Should she stock the same number of hats in each size? Or should she stock more hats in some sizes and fewer in others?

Help Ms. Woods decide. Pretend that she has asked each class in your school to collect and organize data about students' head sizes. She plans to combine the data and then use it to figure out how many hats of each size to stock.

As a class, collect and organize data about one another's head sizes.

1. Ask your partner to help you measure the distance around your head.

   ◆ Wrap the tape measure once around your head.

   ◆ See where the tape touches the end tip of the tape measure.

   ◆ Read the mark where the tape touches the end tip.

   ◆ Read this length to the nearest $\frac{1}{2}$ centimeter.

   Record your head size.   About _____ cm

2. What is the median head size for the class?   About _____ cm

3. Find the following landmarks for the head-size data shown in the bar graph on journal page 47.

   Minimum: _____   Maximum: _____   Range: _____

   Mode: _____   Median: _____

4. How would the landmarks above help Ms. Woods, a clothing store owner, decide how many baseball caps of each size to stock?

   _____

   _____

   _____

   _____

LESSON 2·8

Head Sizes *continued*

Make a bar graph of the head-size data for the class.

title

label

label

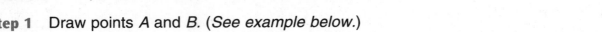

LESSON 2·8 How to Construct a Kite

Follow the steps to construct a kite in the space below.

Step 1   Draw points *A* and *B*. (*See example below.*)

Step 2   Set your compass opening so that it is a little more than half the distance between *A* and *B*. Place the point of the compass on *A* and draw an arc. *Without changing the compass opening,* place the point of the compass on *B* and draw a second arc that intersects the first arc. Label the point where the two arcs meet *C*.

Step 3   Change your compass opening. Set it so that it is almost the full distance between *A* and *B* (about $\frac{4}{5}$ of the width). Repeat Step 2 as shown in the picture, and label the new point of intersection *D*.

Step 4   With your straightedge, connect the 4 points to form a quadrangle.

Example:

Draw your kite here.

D

A •   • B

C

Steps 1–3

D

A   B

C

Step 4

48

LESSON 2·9

Trade-First Subtraction

Write a number model for a ballpark estimate. Solve Problems 1–3 using the trade-first method. Solve Problems 4–6 using any method. Use your ballpark estimates to see if your answers make sense.

1.	2.	3.
$\begin{array}{r} 58 \\ -\ 39 \\ \hline \end{array}$	$\begin{array}{r} 600 \\ -\ 379 \\ \hline \end{array}$	$\begin{array}{r} 2,936 \\ -\ 1,657 \\ \hline \end{array}$
Ballpark estimate: _____	Ballpark estimate: _____	Ballpark estimate: _____
4.	5.	6.
$\begin{array}{r} 73 \\ -\ 37 \\ \hline \end{array}$	$\begin{array}{r} 900 \\ -\ 461 \\ \hline \end{array}$	$\begin{array}{r} 2,468 \\ -\ 1,789 \\ \hline \end{array}$
Ballpark estimate: _____	Ballpark estimate: _____	Ballpark estimate: _____

Try This

7. Name two 3-digit numbers whose difference is 357.

_____ − _____ = 357

LESSON 2·9 **Partial-Differences Subtraction**

Write a number model for a ballpark estimate. Solve Problems 1–3 using the partial-differences method. Solve Problems 4–6 using any method. Use your ballpark estimates to see if your answers make sense.

1.   86
     − 37
	2. 900
	   − 485
		3. 7,584
		  − 2,806

Ballpark estimate: _____	Ballpark estimate: _____	Ballpark estimate: _____

4.   61
     − 26
	5. 400
	   − 271
		6. 3,681
		  − 1,803

Ballpark estimate: _____	Ballpark estimate: _____	Ballpark estimate: _____

Try This

7. Name two 4-digit numbers whose difference is 4,203.

_____ − _____ = 4,203

LESSON 2·9

Math Boxes

---

1. A number has

6 in the tens place,
9 in the millions place,
4 in the thousands place,
3 in the ten-thousands place,
2 in the hundred-thousands place,
0 in the ones place, and
8 in the hundreds place.

Write the number.

___ , ___ ___ ___ , ___ ___ ___

SRB 4

---

2. Write five names for 1,000.

1,000

SRB 149

---

3. Which number sentence is true? Fill in the circle next to the best answer.

(A) $34 - 4 = 31$

(B) $812 < 218$

(C) $423 + 20 > 443$

(D) $123 + 5 + 5 = 113 + 20$

SRB 148 149

---

4. Draw a polygon that has no parallel sides.

SRB 94

---

5. Measure these line segments to the nearest $\frac{1}{2}$ centimeter.

a. _____

   About _____ cm

b. _____

   About _____ cm

SRB 128

---

6. Multiply mentally.

a. _____ $= 5 \times 4$

b. $5 \times 6 =$ _____

c. $9 \times 3 =$ _____

d. $4 \times$ _____ $= 24$

e. $6 \times$ _____ $= 48$

SRB 16

LESSON 2·10 **Math Boxes**

---

1. Divide mentally.

 a. 35 ÷ 5 = _____

 b. 56 ÷ _____ = 8

 c. 32 ÷ _____ = 4

 d. 24 ÷ _____ = 6

 e. 72 ÷ 8 = _____

 f. 40 ÷ 5 = _____

20

---

2. Multiply mentally.

 a. 4 × 7 = _____

 b. 3 × _____ = 15

 c. 7 × _____ = 42

 d. 9 × _____ = 36

 e. 6 × 0 = _____

 f. 1 × 9 = _____

16

---

3. Complete the square facts.

 a. 64 ÷ 8 = _____

 b. 49 ÷ 7 = _____

 c. _____ = 4 × 4

 d. _____ = 3 × 3

 e. 25 ÷ 5 = _____

158

---

4. Tell whether each number sentence is true or false.

 a. 46 + 12 = 53 _____

 b. 36 = 22 + 14 _____

 c. 13 = 84 − 71 _____

 d. 52 − 20 = 34 _____

148

---

5. A grizzly bear can weigh 786 pounds. An American black bear can weigh 227 pounds. What is their combined weight?

_____ pounds

---

6. On average, India produces about 855 movies per year. The United States produces about 762 movies. On average, how many fewer movies per year does the United States produce than India?

_____

Date _____   Time _____

"What's My Rule?"

Complete the "What's My Rule?" tables and state the rules.

1.

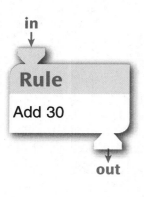

Rule: Add 30

in	out
30	
80	
20	
150	
290	

2.

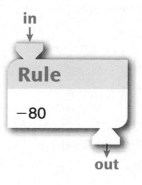

Rule: −80

in	out
	50
	210
	20
	270
	340

3.

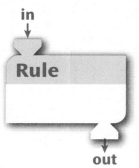

Rule:

in	out
49	72
151	
272	295
	611
	503

4. Rule: There are 12 inches in 1 foot.

in	out
3	36
	60
10	
	264
	720

Try This

5. Rule: _____

in	out
17	−8
12	
27	2
−5	
	0

6. Create your own.

Rule: _____

in	out

LESSON 3·1 A Polygon Alphabet

Try reading this message:

ALL OF THESE LETTERS ARE POLYGONS.

1. Use a straightedge to design a polygon letter for each of the letters shown below. You'll have to simplify, because a polygon can't have any curves, and it can't have any "holes."

   For example, if you look at the letter "P," you see that there is no opening in the upper part. Making it look like this, ▷, would make it easier to read, but it would not be a polygon.

B	C	D
F	M	X

2. Which of the letters you drew are nonconvex (concave) polygons? _____
   How do you know?

   _____

3. Do any of the letters you drew have special names as polygons? Explain.

   _____

   _____

Try This

4. On a separate sheet of paper, design polygon letters for the rest of the uppercase (capital) letters in the alphabet, the 26 lowercase (small) letters, or the 10 digits (0–9).

LESSON 3·1   Math Boxes

1. Write >, <, or = to make each number sentence true.

    a. 1 million _____ 100,000

    b. 73,099 _____ 71,999

    c. 304,608 _____ 304,809

    d. 5,682 _____ 7 hundred

    e. 5,000,236 _____ 5,000,099

SRB 6 149

2. Number of spelling words correct for 10 students on the spelling test:

25, 19, 16, 25, 18, 19, 25, 24, 25, 23

    a. What is the range for this set of numbers? _____

    b. What is the median? _____

SRB 73

3. Make a ballpark estimate. Write a number model to show your strategy.

    a. 3,389 + 2,712

    _____ + _____ = _____

    b. 3,452 − 1,147

    _____ − _____ = _____

SRB 181

4. Complete.

    a. 21 ft = _____ yd

    b. 4 ft = _____ in.

    c. 16 ft = _____ yd _____ ft

    d. 2 yd 2 ft = _____ in.

    e. _____ ft _____ in. = 568 in.

SRB 129

5. Complete.

    a. 7, 15, 23, ____, ____, ____

    Rule: _____

    b. 49, 42, ____, 28, ____, ____

    Rule: _____

    c. ____, ____, 53, 59, ____, 71

    Rule: _____

SRB 160 161

6. Solve mentally or with a paper-and-pencil algorithm.

    a. $3.56
       + $2.49
       ‾‾‾‾‾‾

    b. $6.25
       − $5.01
       ‾‾‾‾‾‾

SRB 34–37

LESSON 3·2 Multiplication/Division Facts Table

*, /	1	2	3	4	5	6	7	8	9	10
1	1	2	3	4	5	6	7	8	9	10
2	2	4	6	8	10	12	14	16	18	20
3	3	6	9	12	15	18	21	24	27	30
4	4	8	12	16	20	24	28	32	36	40
5	5	10	15	20	25	30	35	40	45	50
6	6	12	18	24	30	36	42	48	54	60
7	7	14	21	28	35	42	49	56	63	70
8	8	16	24	32	40	48	56	64	72	80
9	9	18	27	36	45	54	63	72	81	90
10	10	20	30	40	50	60	70	80	90	100

Date _____   Time _____

1. The numbers 28, 35, and 42 are all multiples of __. Circle the best answer.

A   7

B   4

C   6

D   2

SRB 9

2. Complete the "What's My Rule?" table and state the rule.

Rule: _____

in	out
236	331
682	777
	486
938	
647	

SRB 162–166

3. Earth is covered by a rocky outer layer called the *crust,* which is made up of many elements.

a. Is there more aluminum or silicon in Earth's crust?

_____

b. What percentage of Earth's crust is aluminum?

_____

c. Which element makes up most of Earth's crust?

_____

Elements Found in Earth's Crust (percent by weight)

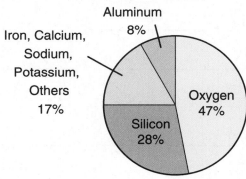

Aluminum 8%

Iron, Calcium, Sodium, Potassium, Others 17%

Oxygen 47%

Silicon 28%

4. Name as many line segments as you can in the figure below.

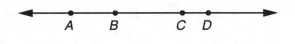

A   B   C D

_____

SRB 90

5. Put these numbers in order from smallest to largest.

0.6   0.06   0.43   0.9

_____

_____

_____

_____

SRB 32 33

57

LESSON 3·3 Patterns in Multiplication Facts

Math Message

Look at the table on journal page 56.

1. Find a pattern in the 9s multiplication facts. Describe the pattern.

_____

_____

_____

_____

_____

2. Find a pattern in the 5s multiplication facts. Describe the pattern.

_____

_____

_____

_____

_____

3. What other patterns can you find in the multiplication facts? Write about some of them.

_____

_____

_____

_____

Math Boxes

---

1. Write >, <, or = to make each number sentence true.

  a. 45,699 _____ 45,609

  b. 67,749 _____ 66,749

  c. 208,775 _____ 200 million

  d. 1,000,000 _____ 858,192

  e. 2 million _____ 20,000,000

SRB
6 149

---

2. Number of days it took 10 students to complete their science projects:

6, 4, 10, 11, 8, 6, 14, 9, 3, 12

  a. What is the range for this set of numbers?

  _____

  b. What is the median?

  _____

SRB
73

---

3. Make a ballpark estimate. Write a number model to show your strategy.

  a. 1,459 + 291

  _____ + _____ = _____

  b. 1,381 − 646

  _____ − _____ = _____

SRB
181

---

4. Complete.

  a. 3 yd = _____ ft

  b. 4 ft = _____ in.

  c. 54 in. = _____ ft _____ in.

  d. $\frac{1}{2}$ yd = _____ ft _____ in.

  e. $17\frac{1}{2}$ yd = _____ in.

SRB
129

---

5. Complete.

  a. 20, 35, 50, _____, _____, _____

  Rule: _____

  b. _____, 68, _____, 94, _____, 120

  Rule: _____

  c. 58, _____, _____, _____, _____, −2

  Rule: _____

SRB
160 161

---

6. Solve mentally or with a paper-and-pencil algorithm.

  a. $10.97
       + $15.60
       _____

  b. $4.56
       − $2.07
       _____

SRB
34–37

LESSON 3·4   **Math Boxes**

---

1. Complete.

a. Name four multiples of 5.

_____, _____, _____, _____

b. Name four multiples of 9.

_____, _____, _____, _____

SRB 9

2. Complete the "What's My Rule?" table and state the rule.

Rule: _____

in	out
502	346
1,238	1,082
	927
871	
1,600	

SRB 162–166

---

3. Mr. Rosario's fourth graders collected data about their favorite fruits.

Favorite Fruits

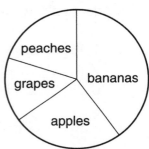

a. Which kind of fruit do most students prefer? _____

b. Which kind is preferred least? _____

c. Do more people prefer apples or peaches? _____

d. About $\frac{1}{4}$ of the students prefer _____.

---

4. Name as many line segments as you can in the figure below.

_____

SRB 90

5. Put these numbers in order from smallest to largest.

0.7   0.007   0.5   0.63

_____

_____

_____

_____

SRB 32 33

LESSON 3·5

Multiplication and Division

SRB
16 20

Equivalents			
3 * 4	12 / 3	12 ÷ 3	3 < 5 (< means "is less than")
3 × 4	$\frac{12}{3}$	3)¯12	5 > 3 (> means "is greater than")

1. Choose 3 Fact Triangles. Write the fact family for each.

_____ * _____ = _____   _____ × _____ = _____   _____ * _____ = _____

_____ * _____ = _____   _____ × _____ = _____   _____ * _____ = _____

_____ / _____ = _____   _____ / _____ = _____   _____ ÷ _____ = _____

_____ / _____ = _____   _____ / _____ = _____   _____ ÷ _____ = _____

2. Solve each division fact.

a. 27 / 3 = _____   Think: How many 3s in 27?

b. _____ = 45 / 5   Think: How many 5s in 45?

c. 36 ÷ 6 = _____   Think: 6 times what number equals 36?

d. 24 / 8 = _____   Think: 8 times what number equals 24?

Try This

3. A cashier has 5 rolls of quarters and 6 rolls of dimes in the cash register. Each roll of quarters is worth $10, and each roll of dimes is worth $5.

a. How much are the rolls of quarters and dimes worth in all? $_____

b. How many quarters are in 1 roll?   _____ quarters

c. How many quarters are in the 5 rolls?   _____ quarters

d. How many dimes are in 1 roll?   _____ dimes

e. How many dimes are in the 6 rolls?   _____ dimes

f. There is also $7.50 worth of half-dollars in the cash register. How many half-dollars is that?   _____ half-dollars

LESSON 3·5

Math Boxes

---

1. Write >, <, or = to make each number sentence true.

   a. 5,389 _____ 3,389

   b. 70,642 _____ 70,699

   c. 6 million _____ 6,000,000

   d. 8,000,032 _____ 8 million, 32 thousand

   e. 400 + 30 + 5 _____ 4,000 + 30 + 5

   6 149

---

2. The number of glasses of milk drunk by 10 students in a week:

   16, 13, 15, 20, 8, 10, 15, 12, 10, 18

   What is the range? Circle the best answer.

   A 8

   B 20

   C 12

   D 14

   73

---

3. Make a ballpark estimate. Write a number model to show your strategy.

   a. 13,685 − 8,379

   _____ − _____ = _____

   b. 7,602 − 3,213

   _____ − _____ = _____

   181

---

4. Complete.

   a. 31 in. = _____ ft _____ in.

   b. 17 ft = _____ yd _____ ft

   c. _____ ft = 14 yd

   d. _____ in. = 2 yd 1 ft

   e. $2\frac{1}{4}$ miles = _____ ft

   129

---

5. Complete.

   a. 7, 11, 15, _____, _____, _____

   Rule: _____

   b. _____, _____, _____, 22, 25, 28

   Rule: _____

   c. _____, 14, _____, 28, _____, 42

   Rule: _____

   160 161

---

6. Solve mentally or with a paper-and-pencil algorithm.

   a. $2.27
      + $4.96

   b. $5.00
      − $3.64

   34–37

Date _____ Time _____

1. Complete the name-collection box.

125

SRB 149

2. A number has

2 in the tens place,
8 in the hundred-thousands place,
5 in the millions place,
7 in the hundreds place,
9 in the ones place,
4 in the thousands place, and
1 in the ten-thousands place.

Write the number:

___ , ___ ___ ___ , ___ ___ ___

SRB 4

3. a. Measure line segment *PQ* to the nearest inch.

P ———————————————————— Q

About _____ inches

b. Measure line segment *RS* to the nearest ½ inch.

R ———————————— S

About _____ inches

SRB 128

4. Solve mentally or with a paper-and-pencil algorithm.

a.   729
       + 432

b.   9,004
       − 515

SRB 10–15

5. Riley estimated the height of his classroom ceiling. Circle the best estimate.

A 7 m

B 3 m

C 20 m

D 15 m

SRB 130

LESSON 3·7

Measuring Air Distances

SRB
145

1. Estimate which city listed below is the closest to Washington, D.C. _____

2. Estimate which city is the farthest. _____

3. Measure the shortest distance between Washington, D.C., and each of the cities shown in the table below. Use the globe scale to convert these measurements to approximate air distances.

   a. Record the globe scale. _____ inch → _____ miles

   b. Complete the table.

Distance from Washington, D.C., to	Measurement on Globe (to the nearest $\frac{1}{2}$ inch)	Air Distance (estimated number of miles)
Cairo, Egypt		
Mexico City, Mexico		
Stockholm, Sweden		
Moscow, Russia		
Tokyo, Japan		
Shanghai, China		
Sydney, Australia		
Warsaw, Poland		
Cape Town, South Africa		
Rio de Janeiro, Brazil		
Choose a city. _____		

4. Explain how you used the globe scale to estimate the air distance between Washington, D.C., and Mexico City, Mexico.

_____

_____

_____

_____

LESSON 3·7 **Math Boxes**

---

1. If 1 centimeter on a map represents 20 kilometers, then

   a. 2 cm represent _____ km.

   b. 5 cm represent _____ km.

   c. 8 cm represent _____ km.

   d. 3.5 cm represent _____ km.

   e. 6.5 cm represent _____ km.

145

---

2. Complete the "What's My Rule?" table and state the rule.

Rule: _____

in	out
40	5
24	3
64	
	4
	9

162–166

---

3. A rock collector has 136 rocks in her collection. She took them to a geologist who said that 57 of them are volcanic. How many of them are not volcanic?

_____

---

4. Solve the riddle. Then use your Geometry Template to draw the shape.

I am a four-sided polygon.

My two short sides are the same length.

My two long sides are the same length.

The sides of the same length are next to each other.

What am I? _____

100

---

5. Make a ballpark estimate. Write a number model to show your strategy.

$2.83 + $0.92 + $3.07 + $7.91

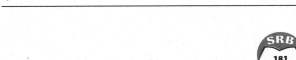

_____

_____

181

---

6. Complete.

   a. 0.1, 0.2, 0.3, _____, _____, _____

   Rule: _____

   b. 1.4, 1.6, 1.8, _____, _____, _____

   Rule: _____

   c. 3, 2.5, 2, _____, _____, _____

   Rule: _____

160 161

LESSON 3·8 Flying to Cairo

Pretend that you are flying from Washington, D.C., to Cairo, Egypt. You have a choice of flying by way of Amsterdam, in the Netherlands, or by way of Rome, Italy. The air distances are shown in the table.

Air Distance between Cities (in miles)

	Amsterdam	Rome
Washington, D.C.	3,851	4,497
Cairo	2,035	1,326

If you fly first class, your ticket will cost $9,250. If you fly economy class, you will save $6,500.

1. About how many more miles is it from Washington, D.C., to Rome than from Washington, D.C., to Amsterdam?

   _____
   (number model)

   Answer: About _____ miles

2. What is the total distance from Washington, D.C., to Cairo by way of Amsterdam?

   _____
   (number model)

   Answer: About _____ miles

3. What is the total distance from Washington, D.C., to Cairo by way of Rome?

   _____
   (number model)

   Answer: About _____ miles

LESSON 3·8 Flying to Cairo *continued*

4. About how many fewer miles will you fly if you go by way of Rome rather than Amsterdam?

_____

(number model)

Answer: About _____ miles

Try This

5. A group of six friends spent $29,500 to fly round-trip from Washington, D.C., to Cairo.

   a. How many members of the group flew first class and how many flew economy?

   Answer: _____ first class and _____ economy

   b. Explain what you did to get your answer.

   _____

   _____

   _____

   _____

6. One of the flights to Amsterdam leaves Washington, D.C., at 1:15 P.M. It lands in Amsterdam 8 hours and 45 minutes later.

   a. At what time does it land, Washington, D.C., time?

   Answer: _____ (A.M. or P.M.?)

   b. At what time does it land, Amsterdam time? (Use the time zones map on pages 276 and 277 in the World Tour section of your *Student Reference Book*.)

   Answer: About _____ (A.M. or P.M.?)

LESSON 3·8 Math Boxes

1. Complete the name-collection box.

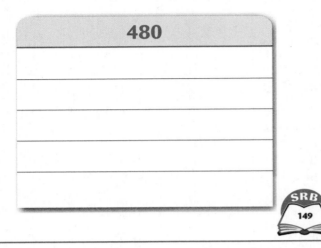

480

SRB
149

2. A number has

1 in the tens place,
3 in the hundreds place,
5 in the ones place,
7 in the hundred-thousands place,
9 in the thousands place,
2 in the ten-thousands place, and
4 in the millions place.

Write the number:

___ , ___ ___ ___ , ___ ___ ___

SRB
4

3. a. Measure the line segment to the nearest inch.

B _____ L

About _____ in.

b. Measure the line segment to the nearest $\frac{1}{2}$ inch.

R _____ T

About _____ in.

SRB
128

4. Solve with a paper-and-pencil algorithm.

a.   604
    + 817

b.   3,005
    − 686

SRB
10–15

5. Sara estimated the length of her arm. Circle the best estimate.

A 5 centimeters

B 50 centimeters

C 100 centimeters

D 150 centimeters

SRB
130

Date _____   Time _____

LESSON 3·9 # Number Sentences

Tell whether each number sentence below is true or false. Write T for true or F for false. If it is not possible to tell, write ? on the answer blank.

1. $7 < 3 + 1$ _____

2. $6 = 36 \div 6$ _____

3. $80 - ? = 40$ _____

4. $28 - 16 = 12$ _____

5. $0 = 4 / 4$ _____

6. $2 * 7$ _____

7. $14 \times 3 < 19 \times 2$ _____

8. $144 + 76 = 880 \div 4$ _____

9. Make up two true number sentences and two false number sentences.

   a. true _____

   b. true _____

   c. false _____

   d. false _____

10. Make up three true number sentences and three false number sentences. Mix them up. Ask your partner to write whether each sentence is true or false.

   Example: $4 * 7 = 34 - 6$ T

   a. _____   _____

   b. _____   _____

   c. _____   _____

   d. _____   _____

   e. _____   _____

   f. _____   _____

69

LESSON 3·9 **Math Boxes**

---

1. If 1 inch on a map represents 30 miles, what would 3 inches represent? Circle the best answer.

  A 10 miles

  B 60 miles

  C 90 miles

  D 300 miles

SRB
145

---

2. Complete the "What's My Rule?" table and state the rule.

Rule: _____

in	out
45	5
81	9
	3
	4
72	

SRB
162–166

---

3. The Statue of Chief Crazy Horse in South Dakota is 563 feet tall. The Statue of Liberty is 151 feet tall. What is the difference in height of the two statues?

_____ feet

---

4. Solve the riddle. Then use your Geometry Template to trace the shape.

I am a polygon.

All my angles have the same measure.

Each of my 5 sides has the same measure.

What am I?

_____

SRB
97

---

5. Make a ballpark estimate. Write a number model to show your strategy.

$2.50 + $0.75 + $3.85 + $12.70

_____

_____

SRB
181

---

6. Complete.

  a. 5.05, 5.06, 5.07, _____, _____, _____

  Rule: _____

  b. 4, 3.8, 3.6, _____, _____, _____

  Rule: _____

  c. 2.7, 3.2, 3.7, _____, _____, _____

  Rule: _____

SRB
160 161

LESSON
3·10   # Parentheses in Number Sentences

SRB
150

Part 1

Make a true sentence by filling in the missing number.

1. **a.** $(30 - 15) * 2 =$ _____   **b.** $30 - (15 * 2) =$ _____

2. **a.** _____ $= 28 / (14 / 2)$ **b.** _____ $= (28 / 14) / 2$

3. **a.** $(6 + 8) / (2 - 1) =$ _____   **b.** $6 + (8 / 2) - 1 =$ _____

Part 2

Make a true sentence by inserting parentheses.

4. **a.** $4 \times 9 - 2 = 34$ **b.** $4 \times 9 - 2 = 28$

5. **a.** $24 = 53 - 11 + 18$ **b.** $60 = 53 - 11 + 18$

6. **a.** $12 / 4 + 2 = 2$ **b.** $12 / 4 + 2 = 5$

7. **a.** $55 = 15 + 10 \times 4$ **b.** $100 = 15 + 10 \times 4$

Try This

Make a true sentence by inserting two sets of parentheses in each problem.

8. **a.** $10 - 4 / 2 * 3 = 24$ **b.** $10 - 4 / 2 * 3 = 1$

Part 3

Pretend you are playing a game of *Name That Number* with only 3 cards per hand.
To name the target number, use all 3 numbers and any operations you want.
For each problem, write a true number sentence containing parentheses using the
3 numbers and the target number.

9. Use: 2, 5, 15   Target number: 5   _____

10. Use: 3, 4, 5   Target number: 17   _____

11. Use: 1, 3, 11   Target number: 4   _____

Date _____   Time _____

1. Complete.

   a. Name all the factors of 50.

   _____, _____, _____, _____, _____, _____

   b. Name the factor pairs of 36.

   _____ and _____

   _____ and _____

   _____ and _____

   _____ and _____

   _____ and _____

   SRB
   7

2. In the 2004 Summer Olympics, which two countries had a combined medal count of 155?

   _____ and _____

Country	Number of Medals
Australia	49
United States	103
China	63
Russia	92

3. What is the mode for the number of books read by the students? Circle the best answer.

   Number of Books Read in a Month

   A   3

   B   4

   C   5

   D   2

   SRB
   73

4. Which of these angles has a measure less than 90 degrees? Circle them.

   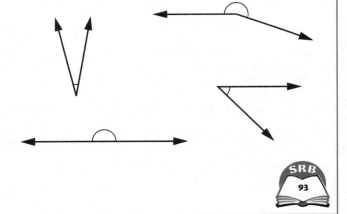

   SRB
   93

5. a. Measure the line segment to the nearest centimeter.

   B ●————————————————————————● N

   About _____ cm

   b. Draw a line segment that is half the length of $\overline{BN}$.

   c. How long is the line segment you drew?   About _____ cm

   SRB
   128

LESSON 3·11   Broken Calculator

Solve each open sentence on your calculator without using the "broken" key.
Only one key is broken in each problem. Record your steps.

1.

Broken Key: ⊖

To Solve: $68 + x = 413$

2.

Broken Key: ⊖

To Solve: $z + 643 = 1{,}210$

3.

Broken Key: ⊕

To Solve: $d - 574 = 1{,}437$

4.

Broken Key: ⊗

To Solve: $w / 15 = 8$

Try This

6. Make up one for your partner to solve.

5.

Broken Key: ÷

To Solve: $s * 48 = 2{,}928$

6.

Broken Key: ☐

To Solve:

LESSON 3·11 Open Sentences

Solve each open sentence. Copy the entire sentence with the solution
in place of the variable. Circle the solution.

1. $48 + d = 70$

$$48 + (22) = 70$$

2. $51 = n + 29$

_____

3. $34 - x = 7$

_____

4. $32 = 76 - p$

_____

5. $h - 6 = 9$

_____

6. $b - 7 = 12$

_____

7. $u - 30 = 10$

_____

8. $5 * m = 35$

_____

9. $y = 3 * 8$

_____

10. $21 / x = 7$

_____

11. $x = 32 / 8$

_____

12. $5 = w / 10$

_____

Try This

13. Mr. O'Connor wrote two open sentences on the board.

$$45 + x = 71$$
$$45 + y = 71$$

Isabel says the two open sentences must have different solutions because
the variables are different.

a. Do you agree with Isabel? _____

b. Explain your answer.

_____

_____

_____

Date _____ Time _____

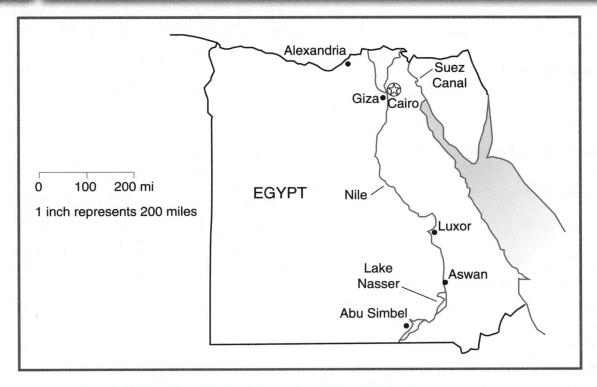

You want to take a trip to Egypt and see the following sights:

♦ Cairo, the capital, on the Nile River, near the Pyramids at Giza

♦ Alexandria, a busy modern city and port on the Mediterranean

♦ The Aswan High Dam across the Nile River, completed in 1970, and Lake Nasser, which formed behind the dam

♦ The temples at Abu Simbel, built more than 3,000 years ago and moved to their present location in the 1960s to escape the rising water of Lake Nasser

You want to know how far it is between locations.

1. The distance between Alexandria and Abu Simbel is about _____ inch(es) on the map.

   That represents about _____ miles.

2. The distance between Cairo and Aswan is about _____ inch(es) on the map.

   That represents about _____ miles.

3. The distance between Abu Simbel and Aswan is about _____ inch(es) on the map.

   That represents about _____ miles.

LESSON
3·11

Math Boxes

1. Complete.

a. Name all the factors of 12.

_____, _____, _____, _____, _____, _____

b. Name the factor pairs of 16.

_____ and _____

_____ and _____

_____ and _____

SRB
7

2. The areas of which two states differ by 944 square miles?

_____ and _____

State	Total Area
Connecticut	5,543 square miles
Rhode Island	1,545 square miles
Delaware	2,489 square miles
New Jersey	8,721 square miles

3. Use the bar graph to answer the questions.

Number of Hours Students Slept Last Night

a. How many students slept 8 hours?

_____

b. What is the mode for the number of hours slept?

_____

SRB
73

4. Which of the angles below have a measure of more than 90 degrees? Circle them.

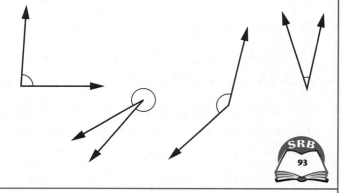

SRB
93

5. a. Measure the line segment to the nearest centimeter.

L •——————————————————————————• P

About _____ cm

b. Draw a line segment that is half the length of $\overline{LP}$.

c. How long is the line segment you drew?   About _____ cm

SRB
128

LESSON 3·12

Math Boxes

---

1. Put these numbers in order from smallest to largest.

0.8   0.08   0.73   0.095

_____

_____

_____

_____

32

---

2. Complete.

a. 0.1, 0.9, 1.7, _____, _____, _____

Rule: _____

b. 5.6, 5.3, 5, _____, _____, _____

Rule: _____

c. 7.23, 7.28, 7.33, _____, _____, _____

Rule: _____

160 161

---

3. Solve mentally or with a paper-and-pencil algorithm.

a. $11.03
       + $4.79

b. $3.56
       − $2.89

34–37

---

4. Blake estimated the length of his little finger. Circle the most reasonable estimate.

A 20 millimeters

B 45 millimeters

C 80 millimeters

D 4 millimeters

130

---

5. a. Measure the line segment to the nearest centimeter.

A •———————————————————————• B

About _____ cm

b. Draw a line segment that is half the length of $\overline{AB}$.

c. How long is the line segment you drew? About _____ cm

128

LESSON 4·1 Place-Value Number Lines

Fill in the missing numbers.

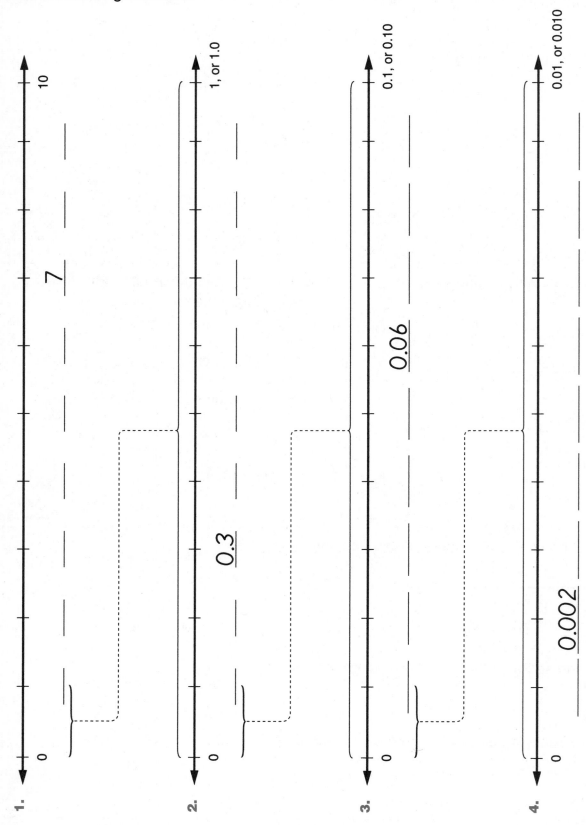

1.   10
     7
     0

2.   1, or 1.0
     0.3
     0

3.   0.1, or 0.10
     0.06
     0

4.   0.01, or 0.010
     0.002
     0

LESSON 4·1 Math Boxes

1. Solve mentally.

    a. $9 * 4 =$ _____

    b. $6 *$ _____ $= 18$

    c. $3 *$ _____ $= 21$

    d. $16 \div 4 =$ _____

    e. $20 \div 4 =$ _____

    f. $54 \div 6 =$ _____

SRB 16 20

2. Solve $199 = p - 408$.
Choose the best answer.

    ◯ $p = 209$

    ◯ $p = 309$

    ◯ $p = 607$

    ◯ $p = 507$

SRB 148

3. In the numeral 9,358,461.72, the 6 is worth 60.

    a. The 4 is worth _____.

    b. The 8 is worth _____.

    c. The 3 is worth _____.

    d. The 9 is worth _____.

    e. The 7 is worth _____.

SRB 30 31

4. Draw and label ray *BY*.
Draw point *A* on it.

SRB 91

5. Insert parentheses to make these number sentences true.

    a. $5 * 4 - 2 = 18$

    b. $25 + 8 * 7 = 81$

    c. $1 = 36 / 6 - 5$

    d. $19 = 15 - 5 + 81 / 9$

SRB 150

6. Estimate the sum. Write a number model to show how you estimated.

$458 + 1,999 + 12,307$

Number model: _____

_____

_____

SRB 181

LESSON 4·2 Tenths and Hundredths

Base-10 Block	Symbol	Value
Flat	□	1
Long	|	$\frac{1}{10}$, or 0.1
Cube	▪	$\frac{1}{100}$, or 0.01

1. Complete the table.

Base-10 Blocks	Fraction Notation	Decimal Notation
‖	$\frac{2}{10}$	0.2
▪▪▪▪▪		
‖‖‖ ‖‖		
□ ‖‖‖ ▪▪▪▪▪		

2. Write each number in decimal notation.

   Example: $\frac{3}{10}$ = ___0.3___

   a. $\frac{4}{10}$ = _____   b. $\frac{71}{100}$ = _____   c. $32\frac{6}{100}$ = _____

3. Draw base-10 blocks to show each number. Draw as few blocks as possible.

   Example: 0.3 ‖‖‖

   a. 0.43   b. 2.16   c. 0.07

4. Write each of the following in decimal notation.

   a. 8 tenths _____   b. 82 hundredths _____   c. 38 and 5 tenths _____

Try This

5. Write 53 thousandths in decimal notation. _____

LESSON 4·2

Math Boxes

1. **a.** What is the maximum number of blocks any student lives from school?

   _____

   b. What is the minimum number of blocks?

   _____

   c. What is the mode? _____

   d. What is the median number of blocks?

   _____

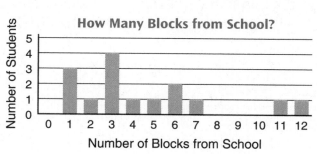

How Many Blocks from School?

Number of Students

Number of Blocks from School

SRB
73

2. Solve mentally or with a paper-and-pencil algorithm.

   a.   391
          + 467

   b.   983
          − 494

SRB
10–15

3. If 1 inch on a map represents 40 miles, then

   a. 2 in. represent _____ mi.

   b. 4 in. represent _____ mi.

   c. 5 in. represent _____ mi.

   d. $2\frac{1}{2}$ in. represent _____ mi.

   e. $1\frac{3}{4}$ in. represent _____ mi.

SRB
145

4. Write as dollars and cents.

   a. 20 dimes = $_____ . _____

   b. 20 nickels = $_____ . _____

   c. 20 quarters = $_____ . _____

   d. 10 quarters and 7 dimes =

   $_____ . _____

5. Solve mentally.

   a. $5 * 3 =$ _____

   b. $5 * 30 =$ _____

   c. _____ $= 5 * 9$

   d. _____ $= 50 * 9$

   e. $6 * 7 =$ _____

   f. $60 * 70 =$ _____

SRB
17

LESSON 4·3 Comparing Decimals

Math Message

1. Arjun thought that 0.3 was less than 0.15. Explain or draw pictures to help Arjun see that 0.3 is more than 0.15.

_____

_____

_____

_____

_____

2. Use base-10 blocks to complete the following table.

> "<" means "is less than."
>
> ">" means "is greater than."

Base-10 Blocks	Decimal	>, <, or =	Decimal	Base-10 Blocks
‖	0.2	>	0.12	|··
·····			0.1	
	0.13			‖|·
‖| ···			0.3	
	1.2			☐☐|
‖|| ▪▪····			0.39	
	2.3			☐||||| |||| |||

Date _____ Time _____

Ordering Decimals

1. Write < or >.

   a. 0.24 _____ 0.18   **b.** 0.05 _____ 0.1   **c.** 0.2 _____ 0.35

   d. 1.03 _____ 0.30   **e.** 3.2 _____ 6.59   **f.** 25.9 _____ 25.72

2. Write your own decimals to make true number sentences.

   a. _____ > _____   **b.** _____ < _____   **c.** _____ < _____

3. Put these numbers in order from smallest to largest.

   a. 0.05, 0.5, 0.55, 5.5   _____ _____ _____ _____
                                       smallest   largest

   b. 0.99, 0.27, 1.8, 2.01   _____ _____ _____ _____
                                       smallest   largest

   c. 2.1, 2.01, 20.1, 20.01   _____ _____ _____ _____
                                       smallest   largest

   d. 0.01, 0.10, 0.11, 0.09   _____ _____ _____ _____
                                       smallest   largest

4. Write your own decimals in order from smallest to largest.

                   _____ _____ _____ _____
                                         smallest   largest

5. "What's green inside, white outside, and hops?"
   To find the answer, put the numbers in order from smallest to largest.

0.66	1	0.2	1.05	0.90	0.01	0.75	0.35	$\frac{25}{100}$	$\frac{50}{100}$	0.05	0.09	5.5
N	I	O	C	W	A	D	S	G	A	F	R	H

   Write your answers in the following table. The first answer is done for you.

0.01												
A												

LESSON 4·3   **Math Boxes**

---

1. Solve mentally.

    a. $5 * \underline{\hspace{1.5cm}} = 40$

    b. $9 * 9 = \underline{\hspace{1.5cm}}$

    c. $9 * \underline{\hspace{1.5cm}} = 27$

    d. $42 \div 6 = \underline{\hspace{1.5cm}}$

    e. $54 \div 9 = \underline{\hspace{1.5cm}}$

    f. $63 \div 7 = \underline{\hspace{1.5cm}}$

SRB 16 20

2. Solve each open sentence.

    a. $100 + w = 175$ $w = \underline{\hspace{1.5cm}}$

    b. $503 + y = 642$ $y = \underline{\hspace{1.5cm}}$

    c. $p + 263 = 319$ $p = \underline{\hspace{1.5cm}}$

    d. $444 - s = 93$ $s = \underline{\hspace{1.5cm}}$

    e. $r - 320 = 600$ $r = \underline{\hspace{1.5cm}}$

SRB 148

---

3. In 34.561

    a. The 3 is worth _____.

    b. The 4 is worth _____.

    c. The 5 is worth _____.

    d. The 6 is worth _____.

    e. The 1 is worth _____.

SRB 30 31

4. Draw and label ray *CT*.
Draw point *A* on it.

SRB 91

---

5. Insert parentheses to make these number sentences true.

    a. $7 * 8 - 6 = 50$

    b. $13 - 4 * 6 = 54$

    c. $10 = 3 + 49 / 7$

    d. $28 = 28 - 6 + 42 / 7$

SRB 150

6. Estimate the sum. Write a number model to show how you estimated.

$3,005 + 9,865 + 2,109$

Number model: _____

_____

_____

SRB 181

LESSON
4·4

A Bicycle Trip

Diego and Alex often take all-day bicycle trips together. During the summer, they took a 3-day bicycle tour. They carried camping gear in their saddlebags for the two nights they would be away from home.

Alex had a **trip meter** that showed miles traveled in tenths of miles. He kept a log of the distances they traveled each day before and after lunch.

Travel Log		
	Distance Traveled	
Timetable	**Before lunch**	**After lunch**
Day 1	27.0 mi	31.3 mi
Day 2	36.6 mi	20.9 mi
Day 3	25.8 mi	27.0 mi

Use estimation to answer the following questions. Do not work the problems out on paper or with a calculator.

1. On which day did they travel the most miles? _____

2. On which day did they travel the fewest miles? _____

3. During the whole trip, did they travel more miles before or after lunch? _____

4. Estimate the total distance they traveled. Choose the best answer.

   ⬭ less than 150 miles   ⬭ between 150 and 180 miles

   ⬭ between 180 and 200 miles   ⬭ more than 200 miles

5. Explain how you solved Problem 4.

   _____

   _____

   _____

6. On Day 1, about how many more miles did they travel after lunch than before lunch?

   _____

7. Diego said that they traveled 1.2 more miles before lunch on Day 1 than on Day 3. Alex disagreed. He said they traveled 2.2 more miles. Who is right? Explain your answer.

   _____

   _____

LESSON 4·4 Math Boxes

1. a. What is the maximum number of movies a student viewed in a month?

_____

b. What is the minimum number of movies? _____

c. What is the mode? _____

d. What is the median? _____

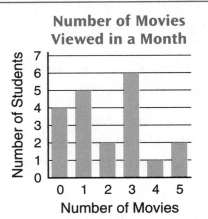

Number of Movies Viewed in a Month

SRB 73

2. Solve mentally or with a paper-and-pencil algorithm.

a. 814
    + 123

b. 754
    − 396

SRB 10–15

3. If 2 centimeters on a map represent 50 kilometers, then

a. 1 cm represents _____ km.

b. 3 cm represent _____ km.

c. 4 cm represent _____ km.

d. 0.5 cm represent _____ km.

e. 8.5 cm represent _____ km.

SRB 145

4. Write 40 quarters and 3 dimes in dollars-and-cents notation. Choose the best answer.

◯ $4.03

◯ $4.30

◯ $10.30

◯ $40.30

5. Solve mentally.

a. $6 * 9 =$ _____

b. $6 * 90 =$ _____

c. _____ $= 5 * 8$

d. _____ $= 50 * 8$

e. $4 * 4 =$ _____

f. $4 * 40 =$ _____

SRB 17

LESSON 4·5 Decimal Addition and Subtraction

Add or subtract mentally or with a paper-and-pencil algorithm.
Pay attention to the + and − symbols.

1. 2.05 + 1.83 = _____

2. 3.04 + 2.8 = _____

3. 2.4 + 3.01 + 0.26 = _____

4. 2.31 − 1.88 = _____

5. 19 + 1.9 = _____

6. 1 − 0.67 = _____

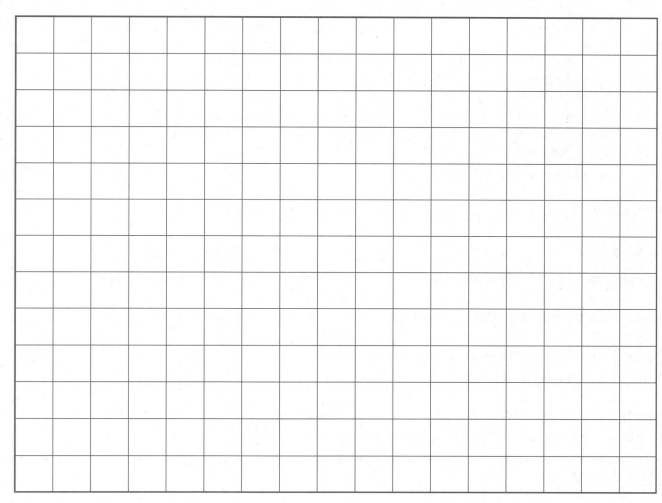

7. Choose one of the problems from above. Explain the method you used
to solve the problem.

_____

_____

_____

LESSON 4·5

Circle Graphs

Percent urban is the number of people out of 100 who live in towns or cities. *Percent rural* is the number of people out of 100 who live in the countryside. Each circle graph below represents the percent of the urban and rural population of an African country.

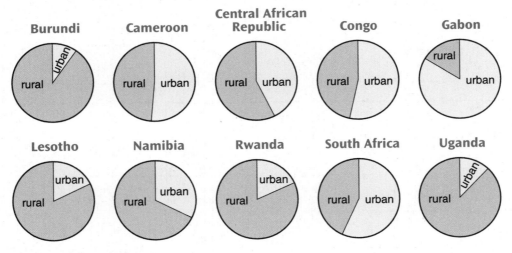

Source: The United Nations

1. For each pair, circle the country with the larger urban population.

   a. Congo   Uganda   b. Rwanda   Gabon

   c. Burundi   South Africa   d. Namibia   Lesotho

2. Which country has the greatest percentage of people living in *urban* areas? _____

3. Which two countries have the greatest percentage
   of people living in *rural* areas?   _____

4. Which two countries have about $\frac{1}{2}$ of their people living
   in urban areas and $\frac{1}{2}$ of their people living in rural areas? _____

Try This

5. Write a question that can be answered from the information in the graphs. Then answer the question.

   Question: _____

   _____

   Answer: _____

Date _____   Time _____

---

1. Insert >, <, or =.

 a. 0.96 _____ 0.4

 b. 0.50 _____ 0.500

 c. 1.3 _____ 1.09

 d. 0.85 _____ 0.86

 e. 0.700 _____ 0.007

SRB
32 33

---

2. a. Measure the length of this line segment to the nearest $\frac{1}{2}$ centimeter.

 _____

 About _____ cm

 b. Draw a line segment 3 centimeters long.

SRB
128

---

3. Fill in the missing numbers.

 a. 28, 35, 42, _____, _____, _____

 Rule: _____

 b. 56, 48, 40, _____, _____, _____

 Rule: _____

 c. 81, _____, 63, _____, 45, _____

 Rule: _____

SRB
160 161

---

4. Solve each open sentence.

 a. $5.9 - T = 5$ $T =$ _____

 b. $9.4 - K = 3$ $K =$ _____

 c. $0.81 - M = 0.43$ $M =$ _____

 d. $F - 2.1 = 6.8$ $F =$ _____

 e. $2.43 = S + 1.06$ $S =$ _____

 f. $R - 12.2 = 4.65$ $R =$ _____

SRB
148

---

5. Add 9 tens, 8 hundredths, and 3 tenths to 34.53.

 What is the result? _____

SRB
36

---

6. Add mentally or with a paper-and-pencil algorithm.

a.	6	**b.**	54
	40		180
	150		240
	+ 1,000		+ 800

SRB
10 11

Date _____ Time _____

LESSON 4·6 Keeping a Bank Balance

Math Message

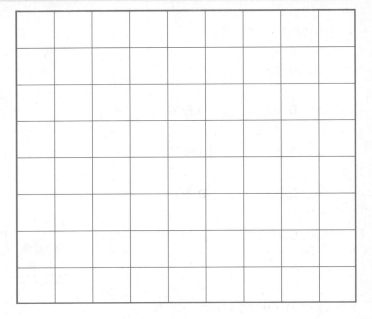

Solve. Show your work on the grid.

1. Cleo went to the store to buy school supplies. She bought a notebook for $2.39, a pen for $0.99, and a set of markers for $3.99. How much money did she spend in all?

   _____

2. Nicholas went to the store with a $20 bill. His groceries cost $13.52. How much change did he get?

   _____

On January 2, Kate's aunt opened a bank account for Kate. Her aunt deposited $100.00 in the account.

Over the next several months, Kate made regular deposits into her account. She deposited part of her allowance and most of the money she made babysitting.

Kate also made a few withdrawals—to buy a radio and some new clothes.

Think about the answers to the following questions:

◆ When you **withdraw** money, do you take money out or put money in?

◆ When you **deposit** money, do you take money out or put money in?

◆ When your money earns **interest,** does this add money to your account or take money away?

The table on the next page shows the transactions (deposits and withdrawals) that Kate made during the first 4 months of the year and the interest she earned.

90

LESSON 4·6 Keeping a Bank Balance *continued*

3. In March, Kate took more money out of her bank account than she put in. In which other month did she withdraw more money than she deposited? _____

4. Estimate whether Kate will have more or less than $100.00 at the end of April. _____

5. Complete the table. Remember to add if Kate makes a deposit or earns interest and to subtract if she makes a withdrawal.

Date	Transaction		Current Balance
January 2	Deposit	$100.00	$ _100.00_
January 14	Deposit	$14.23	+ $ _14.23_ $ _114.23_
February 4	Withdrawal	$16.50	$ _____ $ _____
February 11	Deposit	$33.75	$ _____ $ _____
February 14	Withdrawal	$16.50	$ _____ $ _____
March 19	Deposit	$62.00	$ _____ $ _____
March 30	Withdrawal	$104.26	$ _____ $ _____
March 31	Interest	$0.78	$ _____ $ _____
April 1	Deposit	$70.60	$ _____ $ _____
April 3	Withdrawal	$45.52	$ _____ $ _____
April 28	Withdrawal	$27.91	$ _____ $ _____

LESSON
4·6   **Math Boxes**

1. Solve mentally or with a paper-and-pencil algorithm.

   a. $5.18 − $3.65 = _____   **b.** $16.86 + $9.24 = _____

   c. 0.87 + 0.94 = _____   **d.** 11.2 − 3.9 = _____

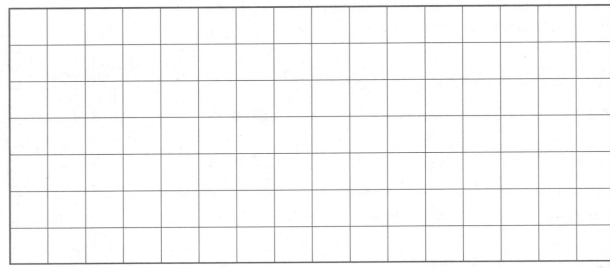

SRB
34–37

---

2. Put these numbers in order from smallest to largest.

   5.92   0.95   9.25   2.95   0.92

   _____ _____ _____ _____ _____

SRB
32 33

3. A trumpeter swan can weigh about 16.8 kilograms. A Manchurian crane can weigh about 14.9 kilograms. How much heavier is a trumpeter swan than a Manchurian crane?

   _____ kilograms

---

4. Number of items students brought to the school food drive:

   28, 26, 3, 8, 2, 6, 8, 13, 1, 5

   What is the

   a. maximum? _____   **b.** minimum? _____

   c. range? _____   **d.** mode? _____

   e. median? _____   **f.** mean? _____

SRB
73 75

5. How do you write the following number using digits: six-hundred million, five thousand, twenty-one? Choose the best answer.

   ⬭ 6,005,210

   ⬭ 600,500,021

   ⬭ 600,500,210

   ⬭ 600,005,021

SRB
4

LESSON 4·7

Math Boxes

---

1. Insert >, <, or =.

a. 0.6 _____ 0.57

b. 0.37 _____ 0.36

c. 2.56 _____ 2.056

d. 0.24 _____ 0.240

e. 0.008 _____ 0.080

SRB 32 33

---

2. a. Measure the length of this line segment to the nearest $\frac{1}{2}$ centimeter.

_____

About _____ cm

b. Draw a line segment 7 centimeters long.

SRB 128

---

3. Fill in the missing numbers.

a. 16, 20, 24, _____, _____, _____

Rule: _____

b. 15, _____, 21, _____, 27, _____

Rule: _____

c. _____, _____, 30, _____, _____, 60

Rule: _____

SRB 160 161

---

4. Solve each open sentence.

a. $D + 1.0 = 9.2$ $D =$ _____

b. $5.6 + K = 10$ $K =$ _____

c. $0.8 - M = 0.6$ $M =$ _____

d. $9.09 + S = 13.64$ $S =$ _____

e. $F + 1.25 = 12.90$ $F =$ _____

f. $R - 0.03 = 1.65$ $R =$ _____

SRB 148

---

5. Add 6 tens, 4 hundredths, and 2 tenths to 367.53.

What is the result? _____

SRB 36

---

6. Add mentally or with a paper-and-pencil algorithm.

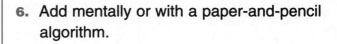

a.	24	b.	6
	80		180
	360		40
	+ 1,200		+ 1,200

SRB 10 11

LESSON 4·7 — Tenths, Hundredths, and Thousandths

Math Message

big cube   flat   long   cube

1. 1 big cube = _____ flats

   1 flat = $\dfrac{1}{10}$ of a big cube

2. 1 big cube = _____ longs

   1 long = $\dfrac{1}{\boxed{}}$ of a big cube

3. 1 big cube = _____ cubes

   1 cube = $\dfrac{1}{\boxed{}}$ of a big cube

Base-10 Block	Symbol	Value
Big Cube	⬓	1
Flat	▢	$\dfrac{1}{10}$, or 0.1
Long	|	$\dfrac{1}{100}$, or 0.01
Cube	▪	$\dfrac{1}{1,000}$, or 0.001

4. Complete the table.

Base-10 Blocks	Fraction Notation	Decimal Notation
▢ ||| ..	$\dfrac{142}{1,000}$	0.142
▢ ▢		
|||| |		
▢ ▢ ▢ ▢ ▢		
⬓ ||| ...		

LESSON 4·7

Tenths, Hundredths, and Thousandths *cont.*

5. Write each number in decimal notation.

Example: $\dfrac{72}{1,000}$ = **0.072**

a. $\dfrac{416}{1,000}$ = _____

b. $\dfrac{17}{1,000}$ = _____

c. $\dfrac{8}{1,000}$ = _____

d. $7\dfrac{9}{100}$ = _____

e. $8\dfrac{14}{1,000}$ = _____

f. $385\dfrac{4}{10}$ = _____

6. Draw base-10 blocks to show each number.

Example: 0.205 ⬜ ⬜ ▸ ▴ ▾ ▴ ▾

a. 0.21

b. 0.306

c. 0.008

d. 1.054

7. Write each of the following in decimal notation.

a. 284 thousandths _____

b. 36 hundredths _____

c. 7 thousandths _____

d. 90 and 16 thousandths _____

e. 15 and 3 tenths _____

f. 408 thousandths _____

8. Write < or >.

a. 0.302 _____ 0.203

b. 0.51 _____ 0.310

c. 0.816 _____ 0.9

d. 1.47 _____ 1.5

e. 0.073 _____ 0.73

f. 4.01 _____ 4.009

Try This

9. Put these numbers in order from smallest to largest: 0.03, 0.009, 0.285, 0.064

_____ _____ _____ _____
    smallest   largest

LESSON 4·8 Measuring Length with Metric Units

1. Your teacher will choose several objects or distances to measure. Measure each to the nearest centimeter. Then compare your measurements with your partner's. If you do not agree, work together to measure the objects again. Record the results in the table.

Object or Distance	My Measurement	Partner's Measurement	Agreed Measurement
	About _____ cm	About _____ cm	About _____ cm
	About _____ cm	About _____ cm	About _____ cm
	About _____ cm	About _____ cm	About _____ cm
	About _____ cm	About _____ cm	About _____ cm
	About _____ cm	About _____ cm	About _____ cm

2. Measure these line segments to the nearest centimeter.

   a. _____

   About _____ centimeters

   b. _____

   About _____ centimeters

3. Measure these line segments to the nearest $\frac{1}{2}$ centimeter.

   a. _____

   About _____ centimeters

   b. _____

   About _____ centimeters

LESSON 4·8

Math Boxes

1. Solve mentally or with a paper-and-pencil algorithm.

   a. 3,309
      + 721

   b. 2,700
      − 1,299

   SRB 10–15

2. Complete.

   a. 1 cm = _____ mm

   b. 5 cm = _____ mm

   c. _____ cm = 30 mm

   d. 100 cm = _____ mm

   e. 200 cm = _____ mm

   SRB 129

3. Tell whether each number sentence is true or false.

   a. 8.77 − 0.08 = 8.50 _____

   b. 35.7 + 22.1 = 57.87 _____

   c. 90.2 − 44.9 < 45 _____

   d. 4.66 + 2.13 > 6 _____

   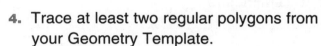SRB 36 37 148

4. Trace at least two regular polygons from your Geometry Template.

   SRB 97

5. Without measuring, estimate the length of your foot from heel to toe. Then measure the length of your foot.

   a. Estimate:

      About _____ cm

   b. Measurement:

      About _____ cm

   SRB 128 130

6. Complete.

   a. Is 47 closer to 40 or 50?

      _____

   b. Name the number halfway between 30 and 40.

      _____

   SRB 182 183

LESSON 4·9 Personal References for Units of Length

Personal References for Metric Units of Length

Use a ruler, meterstick, or tape measure to find common objects that have lengths of 1 centimeter, 1 decimeter, and 1 meter. The lengths do not have to be exact, but they should be close. Ask a friend to look for references with you. You can find more than one reference for each unit. Record the references in the table below.

Unit of Measure	Personal References
1 centimeter (cm)	
1 decimeter (dm), or 10 centimeters	
1 meter (m)	

To be completed in Lesson 5-1.

Personal References for U.S. Customary Units of Length

Use a ruler, yardstick, or tape measure to find common objects that have lengths of 1 inch, 1 foot, and 1 yard. The lengths do not have to be exact, but they should be close. Ask a friend to look for references with you. You can find more than one reference for each unit. Record the references in the table below.

Unit of Measure	Personal References
1 inch (in.)	
1 foot (ft)	
1 yard (yd)	

LESSON
4·9

Measurement Collection for Metric Units of Length

Use your personal references to estimate the length of an object or a distance in centimeters, decimeters, or meters. Describe the object or distance and record your estimate in the table below. Then measure the object or distance and record the actual measurement in the table.

Object or Distance	Estimated Length	Actual Length

LESSON
4·9   **Math Boxes**

---

1. Solve mentally or with a paper-and-pencil algorithm.

   a. $12.63 + $5.66 = _____

   b. $2.46 − $1.34 = _____

   c. 9.6 − 4.8 = _____

   d. 0.64 + 0.47 = _____

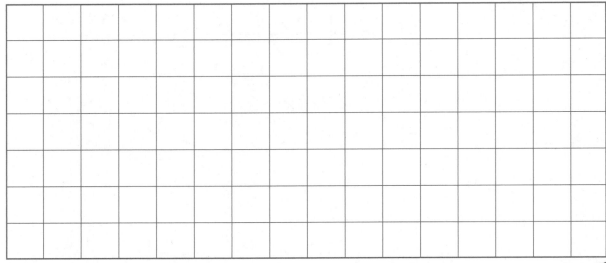

34–37

---

2. Put these numbers in order from smallest to largest.

   1.68   0.78   6.71   0.61

   _____ _____ _____ _____

   32 33

3. The great spotted kiwi bird is about 114.3 centimeters tall. The greater rhea is about 137.1 centimeters tall. How much taller is the greater rhea than the great spotted kiwi bird?

   _____ centimeters

---

4. Make up a set of 7 numbers having the following landmarks:

   mode: 21
   median: 24
   maximum: 35
   range: 20

   ___, ___, ___, ___, ___, ___, ___

   73

5. Write the following numbers using digits:

   a. four hundred eighty-two thousand, one hundred ninety-seven

      _____

   b. eight hundred million, twelve thousand, five

      _____

   4

LESSON 4·10   Measuring in Millimeters

Math Message

On your centimeter ruler, the numbered marks are for centimeters and the little marks between the centimeter marks are for millimeters.

1. Look at your centimeter ruler. How many millimeters are in 1 centimeter? _____ mm

2. Name something that measures about 1 millimeter. _____

   _____

3. Draw a line segment that is 8 centimeters long.

4. Draw a line segment that is 80 millimeters long.

Measure each line segment below using both the millimeter side and the centimeter side of the cm/mm ruler. Record both measurements.

5.   A _____ B

   Length of $\overline{AB}$ = _____ mm = _____ cm

6.   C _____ D   Length of $\overline{CD}$ = _____ mm = _____ cm

7.   E F   Length of $\overline{EF}$ = _____ mm = _____ cm

Measuring Land Invertebrates

An invertebrate is an animal that does not have a backbone. (The backbone is also called the spinal column.) Some invertebrates live on land, others in water. The most common land invertebrates are insects.

The invertebrates shown on page 102, except the earthworm, bumblebee, and mealybug, have been drawn to about actual size. The earthworm can grow to about 4 times the length shown. The bumblebee is shown about twice its actual size and the mealybug about 3 times its actual size.

LESSON 4·10 Measuring Land Invertebrates

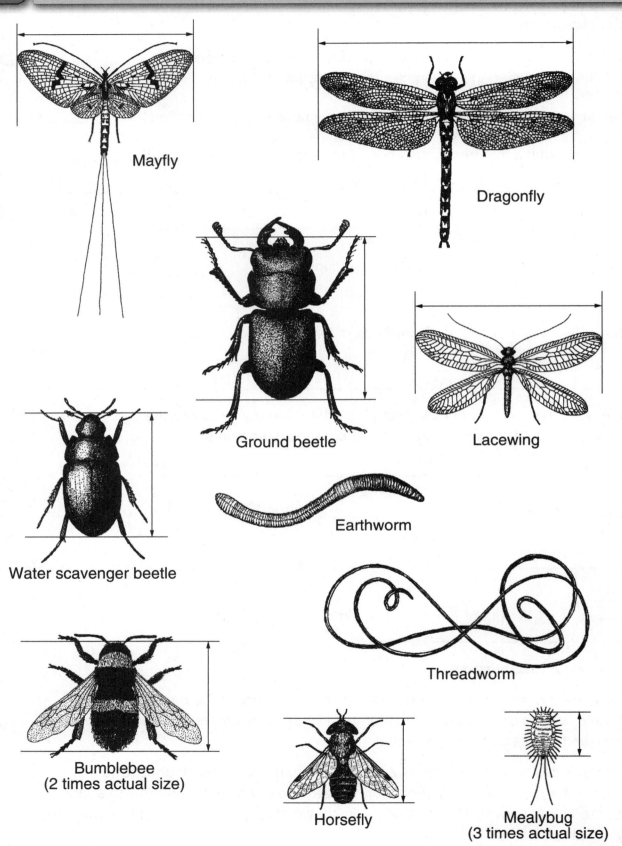

Mayfly

Dragonfly

Ground beetle

Lacewing

Water scavenger beetle

Earthworm

Threadworm

Bumblebee
(2 times actual size)

Horsefly

Mealybug
(3 times actual size)

Date _____   Time _____

Refer to the pictures on page 102
to answer the following questions.

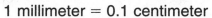 1 centimeter (cm) = 10 millimeters (mm)
1 millimeter = 0.1 centimeter

1. Measure the following invertebrates to the nearest millimeter by finding the
   distance between the two guidelines. Then give the lengths in centimeters.

   a. mayfly   About _____ mm   About _____ cm

   b. dragonfly   About _____ mm   About _____ cm

   c. water scavenger beetle   About _____ mm   About _____ cm

   d. ground beetle   About _____ mm   About _____ cm

   e. lacewing   About _____ mm   About _____ cm

   f. horsefly   About _____ mm   About _____ cm

2. How much longer is the ground beetle than the water scavenger beetle? About _____ cm

3. The bee has been drawn to twice its actual size.
   In reality, which is longer, the bee or the horsefly? _____

   How much longer? About _____ mm

4. The mealybug has been drawn to 3 times its
   actual size. In the space at the right, draw
   a mealybug that is about the actual size.

5. What is the actual size of the mealybug in millimeters? _____ mm

6. How did you solve Problem 5?

   _____

   _____

   _____

7. When straight, the threadworm in the drawing is 306 millimeters long.

   What is its length in centimeters? _____ cm   In meters? _____ m

103

LESSON 4·10 Math Boxes

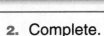

1. Solve mentally or with a paper-and-pencil algorithm.

a. 4,647
   + 3,228

b. 2,500
   − 1,398

SRB
10–15

2. Complete.

a. 7 cm = _____ mm

b. 15 cm = _____ mm

c. 500 cm = _____ m

d. _____ cm = 40 mm

e. _____ cm = 8 m

SRB
129

3. Tell whether each number sentence is true or false.

a. $2.34 - 0.09 = 2.25$ _____

b. $89.6 + 21.7 = 111.3$ _____

c. $56.4 - 23.8 < 33$ _____

d. $5.17 + 3.86 > 10$ _____

SRB
36 37
148

4. Name two properties of a regular polygon.

a. _____

   _____

b. _____

   _____

SRB
97

5. Without measuring, estimate the height of your chair. Then measure it.

a. Estimate:

   About _____ cm

b. Measurement:

   About _____ cm

SRB
128 130

6. Complete.

a. Is 326 closer to 300 or 400?

   _____

b. Name the number halfway between 500 and 800.

   _____

SRB
182 183

LESSON 4·11 **Math Boxes**

---

1. Estimate the sum. Write a number model to show how you estimated.

3,721 + 2,876 + 7,103

Number model: _____

_____

_____

SRB
181

---

2. Solve mentally.

a. 4 * 8 = _____

b. 4 * 80 = _____

c. _____ = 5 * 3

d. _____ = 50 * 3

e. 6 * 6 = _____

f. 6 * 60 = _____

SRB
16 17

---

3. Complete.

a. Is 63 closer to 60 or 70? _____

b. What number is halfway between 80 and 90? _____

c. Is 572 closer to 500 or 600? _____

d. What number is halfway between 300 and 600? _____

SRB
182 183

---

4. Write the following numbers using digits:

a. one million, three hundred forty-six thousand, thirteen

_____

b. twenty-two million, fifteen thousand, three hundred fifty-four

_____

SRB
4

---

5. Add mentally or with a paper-and-pencil algorithm.

a.	**b.**	**c.**	**d.**
35	18	54	48
100	420	180	720
280	120	360	180
+ 800	+ 2,800	+ 1,200	+ 2,700

SRB
10 11

LESSON 5·1 Multiplying Ones by Tens

You can extend a multiplication fact by making one of the factors a multiple of ten.

Example:

Original fact: 2 * 3 = 6

Extended facts: 2 * **30** = _____, or **20** * 3 = _____

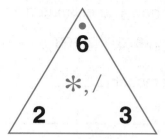

Write a multiplication fact for each Fact Triangle shown below.
Then extend this fact by changing one factor to a multiple of ten.

1.

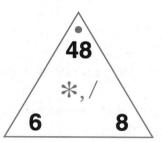

Original fact: _____

Extended fact: _____

2.

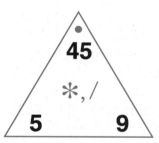

Original fact: _____

Extended fact: _____

3.

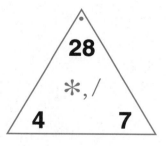

Original fact: _____

Extended fact: _____

4.

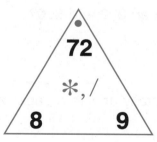

Original fact: _____

Extended fact: _____

5. What shortcut can you use to multiply ones by tens, such as 3 * 60?

_____

_____

Date _____   Time _____

Multiplying Tens by Tens

You can extend a multiplication fact by making both factors multiples of ten.

Example:

Original fact: 3 * 5 = 15

Extended fact: **30 * 50 =** _____

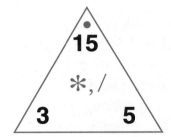

Write a multiplication fact for each Fact Triangle shown below.
Then extend this fact by changing both factors to multiples of ten.

1.

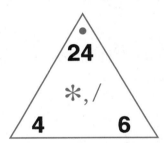

Original fact: _____

Extended fact: _____

2.

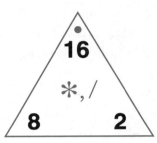

Original fact: _____

Extended fact: _____

3.

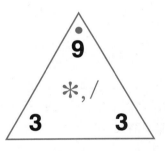

Original fact: _____

Extended fact: _____

4.

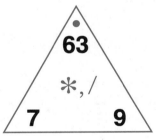

Original fact: _____

Extended fact: _____

5. What shortcut can you use to multiply tens by tens, such as 40 * 60?

_____

_____

Date _____ Time _____

LESSON 5·1 **Math Boxes**

1. A number has

6 in the tenths place,
9 in the hundreds place,
2 in the thousands place,
7 in the ones place,
3 in the tens place, and
5 in the hundredths place.

Write the number.

____ , ____ ____ ____ . ____ ____

SRB
31

2. Solve mentally.

a. $40 * 50 =$ _____

b. $70 * 300 =$ _____

c. $60 *$ _____ $= 180$

d. $90 *$ _____ $= 810$

e. _____ $* 9 = 7,200$

SRB
17

3. Solve mentally or with a paper-and-pencil algorithm.

a.
```
  4,500
    540
    100
+    12
```

b.
```
  2,100
    420
     90
+    18
```

SRB
10 11

4. List all the factors of these numbers.

a. 40 _____

_____

b. 18 _____

_____

c. 28 _____

_____

SRB
7

5. If 1 inch on a map represents 300 miles, then

a. 6 in. → _____ miles

b. 10 in. → _____ miles

c. _____ in. → 900 miles

d. _____ in. → 750 miles

e. $8\frac{1}{2}$ in. → _____ miles

SRB
145

6. a. Five children share 27 tennis balls equally.

Each child gets _____ balls.

There are _____ balls left over.

b. There are 32 cookies for 6 friends.

Each friend gets _____ cookies.

There are _____ cookies left over.

SRB
20

108

LESSON 5·2 Presidential Information

The following table shows the dates on which the most recent presidents of the United States were sworn in and their ages at the time they were sworn in.

SRB
73

President	Date Sworn In	Age
F. D. Roosevelt	March 4, 1933	51
Truman	April 12, 1945	60
Eisenhower	January 20, 1953	62
Kennedy	January 20, 1961	43
Johnson	November 22, 1963	55
Nixon	January 20, 1969	56
Ford	August 9, 1974	61
Carter	January 20, 1977	52
Reagan	January 20, 1981	69
G. H. Bush	January 20, 1989	64
Clinton	January 20, 1993	46
G. W. Bush	January 20, 2001	54

1. What is the median age (the middle age) of the presidents at the time they were sworn in? _____ years

2. What is the range of their ages (the difference between the ages of the oldest and the youngest)? _____ years

3. Who was president for the longest time? _____

4. Who was president for the shortest time? _____

5. Presidents are elected to serve for 1 term. A term lasts 4 years. Which presidents served only 1 term or less than 1 term?

_____

6. Which president was sworn in about 28 years after Roosevelt? _____

7. Roosevelt was born on January 30, 1882. If he were alive today, about how old would he be? _____ years old

LESSON 5·2   Math Boxes

1. Fill in the missing numbers on each number line.

   a.
   0.20   0.30
   ____ ____ ____ ____ ____ ____ ____ ____ ____

   b.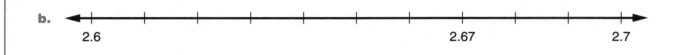
   2.6   2.67   2.7
   ____ ____ ____ ____ ____ ____   ____ ____

2. Measure the line segment to the nearest millimeter. Record the measurement in millimeters and centimeters.

   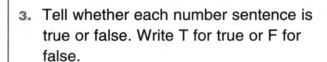

   a. About _____ mm

   b. About _____ cm _____ mm

   c. About _____ . _____ cm

   SRB
   128 129

3. Tell whether each number sentence is true or false. Write T for true or F for false.

   a. 7.89 − 3.36 = 4.53 _____

   b. 12.6 − 4.8 = 3.9 + 3.9 _____

   c. (4.6 + 2.9) − 3.1 < 3.7 _____

   d. 0.20 > 0.68 − (0.42 + 0.11) _____

   SRB
   34–37
   148

4. Insert > or < to make a true number sentence.

   a. 4,500,999 _____ 879,662

   b. 23,468,000 _____ 23,467,000

   c. 568,009,352 _____ 568,010,320

   d. 400,632 _____ 399,800

   SRB
   6

5. Divide mentally.

   a. 45 / 9 = _____

   b. 450 / 90 = _____

   c. 1,000 / 200 = _____

   d. _____ = 2,000 / 40

   e. _____ = 6,300 / 700

   SRB
   21

LESSON 5·3 Math Boxes

---

1. A number has

2 in the hundreds place,
7 in the tenths place,
6 in the hundredths place,
4 in the ones place,
5 in the tens place, and
1 in the thousandths place.

Write the number.

_____ _____ _____ . _____ _____ _____

SRB
31

---

2. Solve mentally.

a. $3 * 40 =$ _____

b. $90 * 70 =$ _____

c. $50 *$ _____ $= 3,000$

d. _____ $* 8 = 4,000$

e. $80 *$ _____ $= 56,000$

SRB
17

---

3. Solve mentally or with a paper-and-pencil algorithm.

a.	72	**b.**	15
	450		240
	160		350
	+ 1,000		+ 5,600

SRB
10 11

---

4. List all the factors of these numbers.

a. 50 _____

_____

b. 32 _____

_____

c. 60 _____

_____

SRB
7

---

5. If 1 centimeter on a map represents 200 miles, what do 4.5 centimeters represent? Fill in the circle next to the best answer.

○ **A.** 850 miles

○ **B.** 900 miles

○ **C.** 450 miles

○ **D.** 800 miles

SRB
43

---

6. a. Sara collected 30 leaves. On the way to school, she lost 2 of them. At school she and her 6 friends shared them equally. How many leaves did each person get?

_____ leaves

b. Ava and her 3 sisters shared 24 mints equally. How many mints did each sister get?

_____ mints

SRB
20

LESSON 5·3 Estimated U.S. Distances and Driving Times

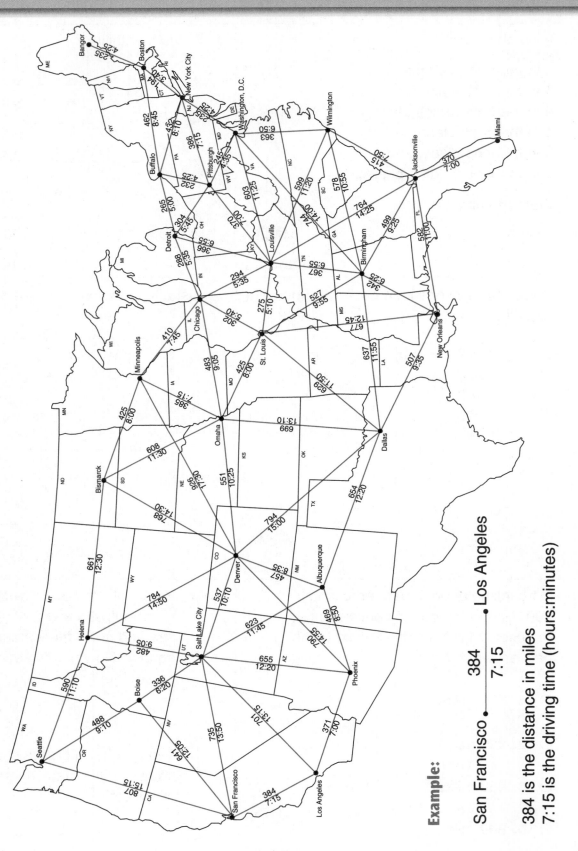

Example:

San Francisco •————384————• Los Angeles
                    7:15

384 is the distance in miles

7:15 is the driving time (hours:minutes)

LESSON 5·3 Planning a Driving Trip

Use the map on journal page 112. Start at your hometown. Plan a driving trip that takes you to 4 other cities on the map. If your hometown is not on the map, find the nearest city on the map to your hometown. Start your driving trip from this city.

Example: Start in Chicago. Drive to St. Louis, Louisville, Birmingham, and then New Orleans.

1. Record your routes, driving distances, and driving times in the table.

From...To	Driving Distance (miles)	Driving Time (hours:minutes)	Rounded Time (hours)
	___ ___ ___	___ : ___	___
	___ ___ ___	___ : ___	___
	___ ___ ___	___ : ___	___
	___ ___ ___	___ : ___	___

2. Estimate how many *miles* you will drive in all.   About _____ miles

3. Estimate how many *hours* you will drive in all.   About _____ hours

4. Tell how many *days* it will take to complete the trip if you plan to drive about 8 hours each day and then stop somewhere for the night.   _____ days

5. Explain how you solved Problems 2–4.

_____

_____

_____

_____

LESSON 5·4 What Do Americans Eat?

The U.S. Department of Agriculture conducts a survey to find out how much food Americans eat. A large number of people are asked to keep lists of all the foods they eat over several days.

These lists are then used to estimate how much of each food was eaten during one year. The average American eats more than 2,000 pounds of food per year. This is about $5\frac{1}{2}$ pounds of food per day.

Results show that the average American eats or drinks about the following amounts in one year:

16	pounds of apples
27	pounds of bananas
5	pounds of broccoli
10	pounds of carrots
30	pounds of cheese
255	eggs
16	pounds of fish
28	cups of yogurt
17	pounds of ice cream
350	cups of milk
22	pounds of candy

Use your answers to the Math Message question to complete these statements.

1. I will eat about _____ eggs in one year.

2. I will drink about _____ cups of milk in one year.

3. I will eat about _____ cups of yogurt in one year.

4. Based on your answers to Problems 1–3, do you think you eat like an average American? Explain why or why not.

_____

_____

LESSON 5·4 | Estimating Averages

◆ Estimate whether the answer will be in the tens, hundreds, thousands, or more.

◆ Write a number model to show how you estimated.

◆ Then circle the box that shows your estimate.

Example: Alice sleeps an average of 9 hours per night. How many hours does she sleep in 1 year?

Number model: $10 * 400 = 4,000$

10s	100s	(1,000s)	10,000s	100,000s	1,000,000s

1. An average of about 23 new species of insects are discovered each day. About how many new species are discovered in one year?

Number model: _____

10s	100s	1,000s	10,000s	100,000s	1,000,000s

2. A housefly beats its wings about 190 times per second. That's about how many times per minute?

Number model: _____

10s	100s	1,000s	10,000s	100,000s	1,000,000s

3. A blue whale weighs about as much as 425,000 kittens. About how many kittens weigh as much as 4 blue whales?

Number model: _____

10s	100s	1,000s	10,000s	100,000s	1,000,000s

4. An average bee can lift about 300 times its own weight. If a 170-pound person were as strong as a bee, about how many pounds could this person lift?

Number model: _____

10s	100s	1,000s	10,000s	100,000s	1,000,000s

LESSON 5·4 **Math Boxes**

---

1. Fill in the missing numbers on each number line.

a.

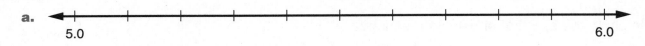

5.0   6.0

_____ _____ _____ _____ _____ _____ _____ _____ _____

b.

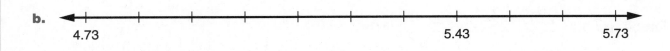

4.73   5.43   5.73

_____ _____ _____ _____ _____ _____   _____ _____

---

2. Measure the line segment to the nearest millimeter. Record the measurement in millimeters and centimeters.

_____

a. About _____ mm

b. About _____ cm _____ mm

c. About _____._____ cm

SRB
128 129

---

3. Which number sentence is true? Fill in the circle next to the best answer.

○ **A.** $(7.6 + 3.8) - 5.2 < 5.9$

○ **B.** $0.50 < 1.43 - (0.77 + 0.19)$

○ **C.** $14.46 - 12.09 = 3.46$

○ **D.** $15.8 - 11.5 = 1.6 + 2.7$

SRB
34–37
148

---

4. Insert > or < to make a true number sentence.

a. 9,000,000 _____ 8,999,999

b. 421,936,500 _____ 422,985,600

c. 68,004,002 _____ 68,004,100

d. 600,523 _____ 601,000

SRB
6

---

5. Divide mentally.

a. 18,000 / 9 = _____

b. 350 / 7 = _____

c. 3,500 / 5 = _____

d. _____ = 5,600 / 800

e. _____ = 42,000 / 60

SRB
21

LESSON 5·5 | Math Boxes

---

1. In 2002, about 32,500,000 people living in the United States had been born in other countries. The circle graph shows where these people were born.

a. Where were most of these people born?

_____

b. About what fraction of the people

were born in Asia? _____

Immigration Statistics

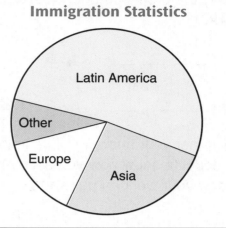

---

2. Estimate the sum. Write a number model to show how you estimated.

a. 387 + 945 + 1,024

Number model:

_____

b. 582 + 1,791 + 2,442

Number model:

_____

181

---

3. Complete.

Rule: _____

in	out
4	280
80	
	490
90	6,300
300	

162–166

---

4. Write two hundred million, three thousand, eighty-eight using digits. Fill in the circle next to the best answer.

○ **A.** 2,030,088

○ **B.** 200,030,088

○ **C.** 20,003,880

○ **D.** 200,003,088

4

---

5. Look at the grid below.

a. In which column is the circle located?

_____

b. In which row is the circle located?

_____

144

LESSON 5·5 The Partial-Products Method

Multiply using the partial-products method. Show your work in the computation grid on page 119.

1. 82 * 3 = _____

2. 6 * 53 = _____

3. 574 * 5 = _____

4. 3 * 470 = _____

5. 2 * 1,523 = _____

6. 3,467 * 3 = _____

Estimate whether your answer will be in the tens, hundreds, thousands, or more. Write a number model to show how you estimated. Circle the correct box. Then calculate the answer. Show your work on page 119.

7. China has the world's longest school year at 251 days. How many school days are in 7 years?

a. Number model: _____

10s	100s	1,000s	10,000s	100,000s	1,000,000s

b. Calculate the answer. _____ days of school

8. People living in the United States eat about 126 pounds of fresh fruit in one year. About how many pounds of fresh fruit would a family of 6 eat in one year?

a. Number model: _____

10s	100s	1,000s	10,000s	100,000s	1,000,000s

b. Calculate the answer. About _____ pounds of fresh fruit

9. Explain how estimation can help you decide whether an answer to a multiplication problem makes sense.

_____

_____

_____

LESSON 5·5 **The Partial-Products Method** *continued*

LESSON 5·5 My Measurement Collection for Units of Length

Use your personal references on journal page 98 to estimate the length or height of an object or distance in inches, feet, or yards. Describe the object or distance, and record your estimate in the table below. Then measure the object or distance, and record the actual measurement in the table.

Object or Distance	Estimated Length	Actual Length

LESSON 5·6 **Math Boxes**

1. a. Measure the line segment to the nearest $\frac{1}{4}$ inch.

   _____

   About _____ inches

   b. Draw a line segment that is half as long as the one above.

   c. How long is the line segment you drew? About _____ inches

   SRB 128

2. Estimate the product. Write a number model to show how you estimated.

   a. 48 * 21

   Number model:

   _____

   b. 98 * 72

   Number model:

   _____

   SRB 184

3. Multiply. Use the partial-products method.

   _____ = 52 * 43

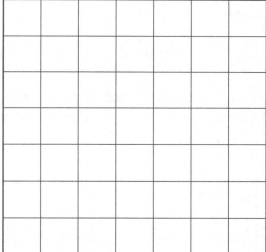

   SRB 18

4. Write each number using digits.

   a. three hundred forty-two thousandths

   _____

   b. six and twenty-five hundredths

   _____

   SRB 27 28

5. If you remove 7 gallons per day from a 65-gallon water tank, how many days will it take to empty the tank?

   _____

   SRB 175

LESSON 5·6

Multiplication Number Stories

Follow these steps for each problem.

a. Decide which two numbers need to be multiplied to give the exact answer. Write the two numbers.

b. Estimate whether the answer will be in the tens, hundreds, thousands, or more. Write a number model for the estimate. Circle the box to show your estimate.

c. On the grid below, find the exact answer by multiplying the two numbers. Write the answer.

1. The average person in the United States drinks about 61 cups of soda per month. About how many cups of soda is that per year?

a. _____ * _____   b. _____   c. _____

numbers that give the exact answer   number model for your estimate   exact answer

10s	100s	1,000s	10,000s	100,000s	1,000,000s

2. Eighteen newborn hummingbirds weigh about 1 ounce. About how many of them does it take to make 1 pound? (1 pound = 16 ounces)

a. _____ * _____   b. _____   c. _____

numbers that give the exact answer   number model for your estimate   exact answer

10s	100s	1,000s	10,000s	100,000s	1,000,000s

Multiplication Number Stories *continued*

3. A test found that a lightbulb lasts an average of 63 days after being turned on.
About how many hours is that?

a. _____ * _____   **b.** _____   **c.** _____

numbers that give   number model for your estimate   exact answer
the exact answer

10s	100s	1,000s	10,000s	100,000s	1,000,000s

4. A full-grown oak tree loses about 78 gallons of water through its leaves per day.
About how many gallons of water is that per year?

a. _____ * _____   **b.** _____   **c.** _____

numbers that give   number model for your estimate   exact answer
the exact answer

10s	100s	1,000s	10,000s	100,000s	1,000,000s

LESSON 5·7 Lattice Multiplication

Use the lattice method to find the products.

1. 3 * 56 = _____

2. 8 * 26 = _____

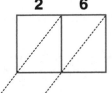

3. 7 * 74 = _____

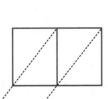

4. 6 * 315 = _____

5. 9 * 284 = _____

6. 47 * 63 = _____

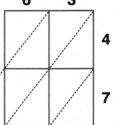

7. 26 * 26 = _____

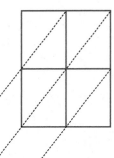

8. 16 * 473 = _____

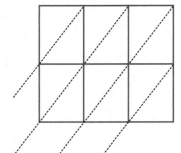

9. 46 * 805 = _____

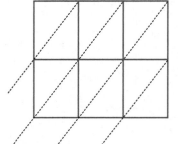

124

**LESSON
5·7**

Math Boxes

1. This is a circle graph of what Seema does on a typical day.

   a. What does she spend the least amount of time doing?

   _____

   b. She spends about the same amount of time at school and doing homework as she does

   _____.

   c. About what fraction of the day does she spend doing chores, eating, and relaxing? _____

   Seema's 24-Hour Day

   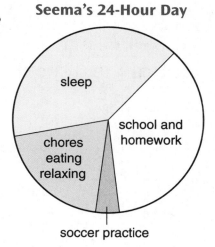

   soccer practice

2. Estimate the sum. Write a number model to show how you estimated.

   a. 715 + 1,904 + 688

   Number model:

   _____

   b. 867 + 2,346 + 3,596

   Number model:

   _____

   181

3. Complete.

   Rule: _____

in	out
50	4,000
70	
	7,200
100	
45	3,600

   162–166

4. Write each number using digits.

   a. twenty-six million, nineteen thousand, eighteen

   _____

   b. three hundred fifty-two million, eight hundred thousand, two hundred

   _____

   4

5. Look at the grid below.

   a. In which column is the triangle located?

   _____

   b. In which row is the triangle located?

   _____

   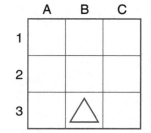

   144

LESSON 5·8 — Reading and Writing Big Numbers

SRB
4

1. Each row in the place-value chart shows a number. Use words to write the name for each number below the chart.

	Billions				Millions				Thousands				Ones		
	100B	10B	1B	,	100M	10M	1M	,	100Th	10Th	1Th	,	100	10	1
a.			7	,	4	0	0	,	0	6	5	,	2	0	0
b.					5	1	,	8	0	0	,	0	0	0	
c.		2	3	,	0	0	0	,	0	0	5	,	1	4	0
d.	1	2	3	,	4	5	6	,	7	8	9	,	0	1	2

a. *7 billion, 400 million, 65 thousand, 200* _____

b. _____

c. _____

d. _____

2. Use digits to write these numbers in the place-value chart below.

   a. 400 thousand, 500

   b. 208 million, 350 thousand, 600

   c. 16 billion, 210 million, 48 thousand, 715

   d. 1 billion, 1 million, 1 thousand, 1

	Billions				Millions				Thousands				Ones		
	100B	10B	1B	,	100M	10M	1M	,	100Th	10Th	1Th	,	100	10	1
a.															
b.															
c.															
d.															

LESSON 5·8

How Much Are a Million and a Billion?

1. How many dots are on the 50-by-40 array page? _____ dots

2. How many dots would be on

   a. 5 pages?   _____ dots

   b. 50 pages?   _____ dots

   c. 500 pages?   _____ dots

3. Each package of paper, or ream, contains 500 sheets. How many dots would be on the paper in

   a. 1 ream? (*Hint:* Look at Problem 2.)   _____ dots

   b. 10 reams? (1 carton)   _____ dots

   c. 100 reams? (10 cartons)   _____ dots

   d. 1,000 reams? (100 cartons)   _____ dots

4. Use digits to write these numbers in the place-value chart below.

   a. 999 thousand   b. 1,000 thousand   c. 999 million   d. 1,000 million

	Billions				Millions				Thousands				Ones		
	100B	10B	1B	,	100M	10M	1M	,	100Th	10Th	1Th	,	100	10	1
a.															
b.															
c.															
d.															

LESSON 5·8 **Internet Users**

The table below shows the estimated number of people who used the Internet in several different countries in 2004.

Internet Users in 2004	
Country	Users
France	25,470,000
Greece	2,710,000
Hungary	2,940,000
Italy	25,530,000
Poland	10,400,000
Spain	13,440,000

Source: The World Factbook

1. Which of these countries had the most Internet users? _____

2. Which of these countries had the fewest Internet users? _____

3. About how many more Internet users were there in France than in Spain?

   Number model: _____

   Answer: _____

4. Write true or false.

   a. There were more Internet users in Greece, Poland, and Hungary combined than in Spain. _____

   b. There were about 6 times as many Internet users in France as there were in Hungary. _____

5. Why do you think the Internet usage data is rounded to the nearest 10,000 instead of an actual count?

   _____

   _____

6. The United States had about 62 times as many Internet users as Hungary. About how many Internet users were there in the United States?

   Number model: _____

   Answer: _____

LESSON 5·8   **Math Boxes**

---

1. a. Measure the line segment to the nearest $\frac{1}{4}$ inch.

R _____ S

About _____ inches

b. Draw a line segment that is half as long as the one above.

c. How long is the line segment you drew?   About _____ inches

128

---

2. Estimate the product. Write a number model to show how you estimated.

a. 41 * 83

Number model:

_____

b. 75 * 32

Number model:

_____

184

3. Multiply. Use the partial-products method.

_____ = 46 * 98

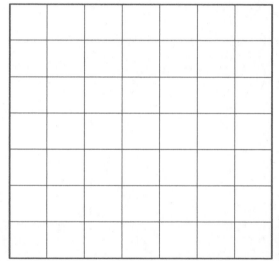

18

---

4. Write each number using digits.

a. seven hundred six thousandths

_____

b. three and four hundredths

_____

27 28

5. You have 75 apples. You need 8 apples to make a pie. How many pies can you bake?

○ **A.** 9

○ **B.** 8

○ **C.** 7

○ **D.** 6

175

LESSON 5·9 Place Value and Powers of 10

Millions	Hundred Thousands	Ten Thousands	Thousands	Hundreds	Tens	Ones
1,000,000			1,000	100		1
10 [100,000s]			10 [100s]	10 [10s]		10 [tenths]
		10 * 10 * 10 * 10		10 * 10		
	10^5		10^3	10^2		10^0

Fill in this place-value chart as follows:

1. Write standard numbers in Row 1.

2. In Row 2, write the value of each place to show that it is 10 times the value of the place to its right.

3. In Row 3, write the place values as products of 10s.

4. In Row 4, show the values as powers of 10. Use exponents. The exponent shows how many times 10 is used as a factor. It also shows how many zeros are in the standard number.

Date _____   Time _____

1. Estimate the sum. Write a number model to show how you estimated.

a. 1,254 + 8,902 + 2,877

Number model:

_____

b. 12,645 + 7,302 + 15,297

Number model:

_____

SRB 181

2. Which number sentence is true? Fill in the circle next to the best answer.

○ **A.** 5,800,000 = 58 million

○ **B.** 62 million > 3,100,000,000

○ **C.** 10^3 = 1,000

○ **D.** 100,000 = 10^2

SRB 5 6

3. Draw lines to match each word to the correct pair or pairs of line segments.

perpendicular

parallel

intersecting

SRB 94 95

4. Complete.

Rule: _____

in	out
1.29	1.36
6.47	
	5.17
12.66	
7.93	8.00

SRB 162–166

5. Multiply. Use a paper-and-pencil algorithm.

9 * 258 = _____

SRB 18 19

6. Which of the angles below has a measure of about 90 degrees? Circle it.

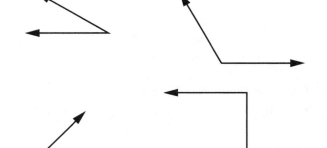

SRB 92 93

LESSON 5·10 Evaluating Large Numbers

1. Round the attendance figures in the table to the nearest hundred-thousand.

2004 Major League Baseball Home Game Attendance for 10 Teams

Team	Total Home Game Attendance	Attendance Rounded to the Nearest 100,000
Atlanta Braves	2,603,484	
Baltimore Orioles	2,682,917	
Boston Red Sox	2,650,063	
Cleveland Indians	2,616,940	
Colorado Rockies	2,737,838	
Los Angeles Dodgers	3,131,255	
New York Yankees	3,465,807	
Seattle Mariners	3,540,658	
St. Louis Cardinals	3,011,216	
Texas Rangers	2,352,397	

Source: Information Please—Baseball Attendance

2. How do you think attendance figures for major league baseball games are obtained?

_____

_____

_____

3. Do you think *exactly* 2,737,838 people were at the home games played by the Colorado Rockies? Explain your answer.

_____

_____

_____

4. You rounded the figures in the table above to the nearest hundred-thousand. Which teams have the same attendance figures based on these rounded numbers?

_____

_____

LESSON 5·10   **Math Boxes**

1. **a.** Measure the line segment to the nearest $\frac{1}{4}$ inch.

   _____
   T   *G*

   About _____ inches

   b. Draw a line segment that is half as long as the one above.

   c. How long is the line segment you drew?   About _____ inches

   SRB 128

2. Estimate the product. Write a number model to show how you estimated.

   a. 37 * 91

   Number model:

   _____

   b. 53 * 17

   Number model:

   _____

   SRB 184

3. Multiply. Use the partial-products method.

   _____ = 43 * 89

   SRB 18

4. Write fourteen and three-tenths using digits. Fill in the circle next to the best answer.

   ○ **A.** 14.03

   ○ **B.** 14.003

   ○ **C.** 14.310

   ○ **D.** 14.3

   SRB 27 28

5. There are 60 trading cards. Each student gets 5 cards. How many students get trading cards?

   _____ students

   SRB 175

LESSON 5·11 Traveling to Europe

It is time to leave Africa. Your destination is Region 2—the continent of Europe. You and your classmates will fly from Cairo, Egypt to Budapest, Hungary. Before exploring Hungary, you will collect information about the countries in Region 2. You may even decide to visit another country in Europe after your stay in Budapest.

Use the World Tour section of your *Student Reference Book* to answer the questions.

1. Which country in Region 2 has

   a. the largest population?   _____
                                country   population

   b. the smallest population? _____
                                country   population

   c. the largest area?   _____
                                country   area

   d. the smallest area?   _____
                                country   area

Use the Climate and Elevation of Capital Cities table on page 297.

2. From December to February, which capital in Region 2 has

   a. the warmest weather?   _____
                                capital   country   temperatures

   b. the coolest weather?   _____
                                capital   country   temperatures

   c. the greatest amount of rain? _____
                                capital   country   inches rainfall

   d. the least amount of rain?   _____
                                capital   country   inches rainfall

Use the Population Data table on page 301.

3. Which country in Region 2 has

   a. the greatest percent of population ages 0–14?   _____
                                                       country   percent

   b. the smallest percent of population ages 0–14?   _____
                                                       country   percent

Date _____   Time _____

LESSON 5·11

Math Boxes

---

1. Estimate the sum. Write a number model to show how you estimated.

a. 799 + 11,304 + 48,609

Number model:

_____

b. 4,382 + 6,911 + 7,035

Number model:

_____

181

2. Write <, >, or = to make each number sentence true.

a. 356,789 _____ 354,999

b. 670,000 _____ 67,000,000

c. 62 million _____ 9,700,000

d. 105,000,000 _____ 15,500,000

5   6

e. 10^4 _____ 1,000

---

3. a. Draw a pair of parallel line segments.

b. Draw a pair of intersecting line segments.

94   95

4. Complete.

Rule: _____

in	out
6.46	6.58
3.08	
	11.34
25.25	
	100.1
63.09	63.21

162–166

---

5. Multiply. Use a paper-and-pencil algorithm.

7 * 208 = _____

18   19

6. Which of the angles below has a measure less than 90 degrees? Circle it.

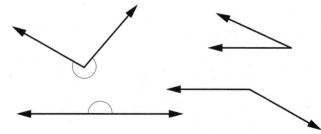

93

LESSON 5·12 Math Boxes

1. There are 240 chairs to set up for the concert. Each row has 40 chairs in it. How many rows are there?

_____ rows

SRB 21

2. The senior class at Rees High School raised $1,895 for five charities in the community. The money will be shared equally. How much money will each charity receive?

_____

SRB 22 23

3. Look at the grid below.

a. In which column is the star located?

_____

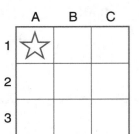

	A	B	C
1	☆		
2			
3			

b. In which row is the star located?

_____

SRB 144

4. Draw a right angle with vertex *K*.

SRB 93

5. Divide.

a. 72 / 9 = _____

b. 720 / 90 = _____

c. 7,200 / 900 = _____

d. _____ = 42 / 7

e. _____ = 420 / 7

f. _____ = 4,200 / 7

g. 28 / 4 = _____

h. 28,000 / 40 = _____

i. 28,000 / 400 = _____

j. _____ = 24 / 6

k. _____ = 2,400 / 60

l. _____ = 24,000 / 6

SRB 20 21

LESSON 6·1   Math Boxes

1. Measure each line segment to the nearest millimeter.

a. _____

    R   S

About _____ cm   _____ mm

b. _____

    C   S

About _____ cm   _____ mm

2. Round 409,381,886 to the nearest

a. hundred.

_____

b. ten thousand.

_____

c. ten million.

_____

d. hundred million.

_____

3. Multiply. Use a paper-and-pencil method.

_____ = 86 * 29

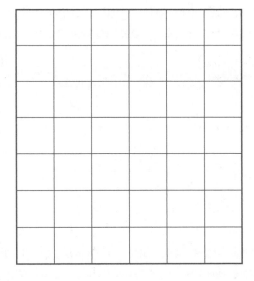

4. Complete.

a. $10^2 =$ _____

b. $10^{\boxed{}} = 10 * 10 * 10 * 10$

c. $1,000 = 10^{\boxed{}}$

d. 10 to the ninth power =

_____

5. Circle $\frac{5}{6}$ of the squares.

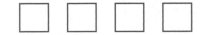

LESSON 6·1 Multiplication/Division Number Stories

Fill in each Multiplication/Division Diagram. Then write a number model.
Be sure to include a unit with your answer.

1. The profit from the Maple Street lemonade stand was $36. Four girls will share this amount equally. What will each girl's share be?

girls	dollars per girl	dollars in all

Number model: _____

Answer: _____

2. Sheila has 57 pictures to put in her photo album. She can put 6 pictures on each page. How many pages will be completely filled up when she is finished?

pages	pictures per page	pictures in all

Number model: _____

Answer: _____

3. Reuben walks a total of 35 blocks going to and from school each week. He always walks the same route. How many blocks does he walk each day?

days	blocks per day	blocks in all

Number model: _____

Answer: _____

LESSON 6·1 **Multiplication/Division Number Stories** *cont.*

4. Hassan is helping his teacher put 8 centimeter cubes into each paper cup for a math project. How many paper cups can he fill if there are 79 cubes?

paper cups	cubes per paper cup	cubes in all

Number model: _____

Answer: _____

Try This

5. Mr. Henning's fourth-grade class is planning a field trip to see a play. The bus will cost $100, and the tickets will cost $125. The 25 students will share the total cost equally. How much will each student pay for the field trip?

_____	_____ per _____	_____ in all

Number model: _____

Answer: _____

6. Martina and 3 friends sold 636 boxes of cookies for their club. Each girl sold the same number of boxes. How many boxes of cookies did each girl sell?

_____	_____ per _____	_____ in all

Number model: _____

Answer: _____

LESSON 6·1 | Extended Multiplication Facts

SRB
17

1. 9 * 5 = _____

9 * 50 = _____

90 * 5 = _____

90 * 50 = _____

900 * 5 = _____

90 * 500 = _____

2. 8 * 7 = _____

8 * 70 = _____

80 * 7 = _____

80 * 70 = _____

800 * 7 = _____

80 * 700 = _____

3. 4 * 9 = _____

4 * 90 = _____

40 * 9 = _____

40 * 90 = _____

400 * 9 = _____

40 * 900 = _____

4. 6 * _____ = 18

60 * _____ = 180

60 * _____ = 1,800

_____ * 60 = 180

_____ * 600 = 1,800

30 * _____ = 18,000

5. _____ * 8 = 48

_____ * 80 = 480

_____ * 80 = 4,800

60 * _____ = 480

6 * _____ = 4,800

6 * _____ = 48,000

6. 8 * _____ = 24

8 * _____ = 2,400

80 * _____ = 2,400

_____ * 30 = 240

_____ * 3 = 240

_____ * 300 = 240,000

Math Boxes

1. There are 32 students in the class. A yearbook page can show 8 student photos. How many pages are needed to include all the student photos?

pages	photos per page	photos in all

Number model: _____

Answer: _____ pages

2. Solve each open sentence.

a. $24 = a * (5 + 1)$ $a =$ _____

b. $54 / 6 = 81 / b$ $b =$ _____

c. $(c + 4) / 3 = 7$ $c =$ _____

d. $m - 3.87 = 7.49$ $m =$ _____

e. $0.98 + 4.83 = f + 4.35$ $f =$ _____

SRB
148

3. Use a paper-and-pencil algorithm to add or subtract.

a.   0.85
  $+ 0.53$

b.   0.64
  $+ 1.73$

c.   12.38
  $- 1.09$

d.   3.05
  $- 0.67$

SRB
34–37

4. Complete.

a. 670 cm = _____ m

b. 4,800 cm = _____ m

c. 916 cm = _____ m _____ cm

d. 18 m = _____ cm

SRB
129

5. Name a fraction equivalent to $\frac{1}{2}$. Circle the best answer.

A. $\frac{3}{4}$

B. $\frac{8}{9}$

C. $\frac{5}{10}$

D. $\frac{3}{5}$

SRB
51

LESSON 6·2 **Solving Division Problems**

For Problems 1–6, fill in the multiples-of-10 list if it is helpful. If you prefer to solve the division problems in another way, show your work.

1. José's class baked 64 cookies for the school bake sale. Students put 4 cookies in each bag. How many bags of 4 cookies did they make?

    10 [4s] = _____   Number model: _____

    20 [4s] = _____   Answer: _____ bags

    30 [4s] = _____

    40 [4s] = _____

    50 [4s] = _____

2. The community center bought 276 cans of soda for a picnic. How many 6-packs is that?

    10 [6s] = _____   Number model: _____

    20 [6s] = _____   Answer: _____ 6-packs

    30 [6s] = _____

    40 [6s] = _____

    50 [6s] = _____

3. Each lunch table at Johnson Elementary School seats 5 people. How many tables are needed to seat 191 people?

    10 [5s] = _____   Number model: _____

    20 [5s] = _____   Answer: _____ tables

    30 [5s] = _____

    40 [5s] = _____

    50 [5s] = _____

LESSON 6·2

Solving Division Problems *continued*

4. The preschool held a tricycle parade. Trent counted 135 wheels.
How many tricycles is that?

10 [3s] = _____

20 [3s] = _____

30 [3s] = _____

40 [3s] = _____

50 [3s] = _____

Number model: _____

Answer: _____ tricycles

---

5. How many 8s are there in 248?

10 [8s] = _____

20 [8s] = _____

30 [8s] = _____

40 [8s] = _____

50 [8s] = _____

Number model: _____

Answer: _____

---

6. How many 7s are in 265?

10 [7s] = _____

20 [7s] = _____

30 [7s] = _____

40 [7s] = _____

50 [7s] = _____

Number model: _____

Answer: _____

LESSON 6·3   Partial-Quotients Division Algorithm

These notations for division are equivalent:

$$6\overline{)134} \qquad\qquad 134 \div 6 \qquad\qquad 134\ /\ 6 \qquad\qquad \frac{134}{6}$$

1. There are 6 pencils in each pack. How many packs can be made from 96 pencils?

Number model: _____

Answer: _____ packs

How many pencils are left over? _____ pencils

2. Phil has $79 to purchase books. Books cost $7 each. How many books can Phil buy?

Number model: _____

Answer: _____ books

How many dollars are left over? _____ dollars

3. There are 184 plants to be put into pots. Each pot can hold 8 plants. How many pots are needed?

Number model: _____

Answer: _____ pots

How many plants are left over? _____ plants

4. The principal shared 395 cookies equally among 9 classes. How many cookies did each class receive?

Number model: _____

Answer: _____ cookies

How many cookies were left over? _____ cookies

LESSON 6·3 Partial-Quotients Division Algorithm *cont.*

Divide.

5. 3)‾87‾

Answer: _____

6. 331 ÷ 7

Answer: _____

Try This

7. Twelve shirts fit into a box. There are 372 shirts to be put into boxes. How many boxes are needed?

Number model: _____

Answer: _____ boxes

How many shirts are left over? _____ shirts

8. There are _____ players in the league. (Write a number greater than 100.)

There are _____ players on each team. (Write a number between 3 and 9.)

How many teams can be made?

Number model: _____

Answer: _____ teams

How many players are left over? _____ players

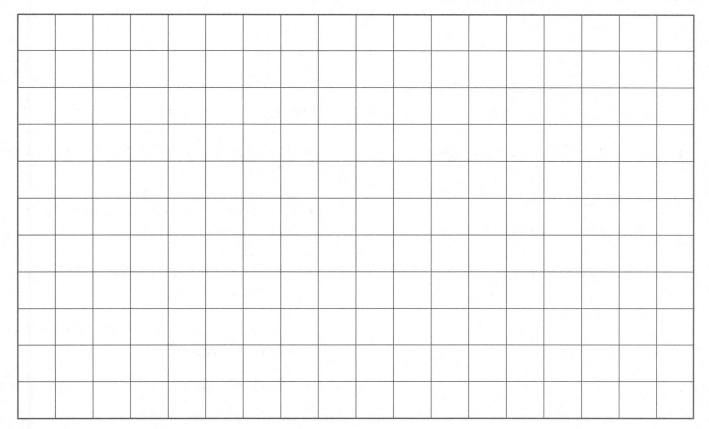

**LESSON
6·3** **Place Value in Decimals**

1. Write these numbers in order from smallest to largest.

1.26   0.58   1.09   1.091   0.35

_____ smallest

_____

_____

_____

_____ largest

2. A number has

6 in the tenths place,
4 in the ones place,
5 in the hundredths place, and
9 in the tens place.

Write the number.

_____ _____ . _____ _____

3. Write the smallest number you can make with the following digits:

3   6   4   7   2

_____

4. What is the value of the digit 4 in the numerals below?

a. 37.48 _____

b. 49.08 _____

c. 0.942 _____

d. 1.664 _____

5. Write each number using digits.

a. four and seventy-two hundredths

_____

b. nine hundred thirty-five thousandths

_____

6. I am a four-digit number less than 10.

◆ The digit in the tenths place is the result of dividing 36 by 4.

◆ The digit in the hundredths place is the result of dividing 42 by 7.

◆ The digit in the ones place is the result of dividing 72 by 8.

◆ The digit in the thousandths place is the result of dividing 35 by 5.

What number am I?

_____ . _____ _____ _____

LESSON 6·3   Math Boxes

1. Measure each line segment to the nearest millimeter.

   a. _____

     P   S

     About _____ cm _____ mm

   b. _____

     A   B

     About _____ cm _____ mm

SRB
128

2. Round 5,906,245 to the nearest

   a. million.

     _____

   b. ten thousand.

     _____

   c. thousand.

     _____

   d. hundred.

     _____

SRB
182   183

3. Multiply. Use a paper-and-pencil method.

     _____ = 58 * 52

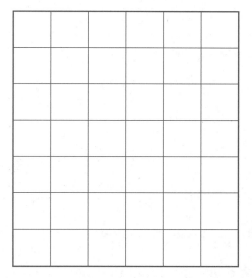

SRB
18   19

4. Complete.

   a. 10^4 = _____

   b. $10^{\square}$ = 10 * 10 * 10 * 10 * 10

   c. 100 = $10^{\square}$

   d. 10 to the seventh power =

     _____

SRB
5

5. Circle $\frac{1}{2}$ of the squares.

SRB
44

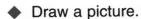

Interpreting Remainders

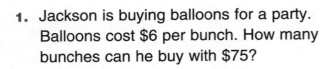

SRB
179

For each number story:

◆ Draw a picture.
◆ Write a number model.
◆ Use a division algorithm to solve the problem.
◆ Decide what to do about the remainder.

1. Jackson is buying balloons for a party. Balloons cost $6 per bunch. How many bunches can he buy with $75?

   Picture:

   Number model: _____

   Answer: _____ bunches

   What did you do about the remainder? Circle the answer.

   A. Ignored it

   B. Reported it as a fraction or decimal

   C. Rounded the answer up

   Why? _____

   _____

   _____

2. Rosa is buying boxes to hold all 128 of her CDs. Each box holds 5 CDs. How many boxes are needed to store all of her CDs?

   Picture:

   Number model: _____

   Answer: _____ boxes

   What did you do about the remainder? Circle the answer.

   A. Ignored it

   B. Reported it as a fraction or decimal

   C. Rounded the answer up

   Why? _____

   _____

   _____

LESSON 6·4 | **Interpreting Remainders** *continued*

3. Lateefah won 188 candy bars in a raffle. She decided to share them equally with 7 of her classmates and herself. How many candy bars did each person receive?

Picture:

Number model: _____

Answer: _____ candy bars

What did you do about the remainder? Circle the answer.

A. Ignored it

B. Reported it as a fraction or decimal

C. Rounded the answer up

Why? _____

_____

_____

_____

Try This

Write each answer as a mixed number by rewriting the remainder as a fraction.

4. 2)27 13 R1   _____

5. 10)883 88 R3   _____

6. 16)252 15 R12   _____

Write each answer as a decimal.

7. 39 ÷ 2 = 19 R1   _____

8. 183 ÷ 12 = 15 R3   _____

9. 2,067 ÷ 5 = 413 R2   _____

LESSON 6·4

Math Boxes

---

1. Joe ordered 72 plants for his patio garden. Each pot holds 4 plants. How many pots are needed to hold all of the plants?

pots	plants per pot	plants in all

Number model: _____

Answer: _____

2. Solve each open sentence.

a. $(6 + 9) + (3 * A) = 30$ $A =$ _____

b. $24 ÷ 8 = 21 ÷ B$ $B =$ _____

c. $72 = (2 * C) * 9$ $C =$ _____

d. $6.2 + 0.79 = D$ $D =$ _____

e. $8.91 - E = 2.72$ $E =$ _____

SRB
35–37
148

---

3. Use a paper-and-pencil algorithm to add or subtract.

a.
```
   0.37
+ 0.26
```

b.
```
   2.9
+ 5.01
```

c.
```
   6.79
- 6.55
```

d.
```
   7.80
- 3.65
```

SRB
34–37

---

4. How many centimeters are in 12 meters? Circle the best answer.

A. 0.12

B. 1.2

C. 120

D. 1,200

SRB
129

5. Circle the fractions equivalent to $\frac{1}{2}$.

$\frac{8}{16}$ $\frac{5}{6}$ $\frac{6}{12}$

$\frac{2}{3}$ $\frac{12}{24}$ $\frac{8}{15}$

SRB
51

LESSON 6·5 Math Boxes

1. Insert parentheses to make each number sentence true.

   a. $15 + 5 * 6 = 120$

   b. $7 + 9 * 2 = 25$

   c. $77 = 1 + 6 * 6 + 5$

SRB
150

2. Draw a line segment that is 2 inches long. Mark and label the following inch measurements on the line segment:

$$\frac{1}{2}, \frac{3}{4}, 1, 1\frac{1}{2}, \text{ and } 2$$

SRB
128

3. The Sports Boosters raised $908 at their annual chili supper. Four athletic teams will share the money equally.

How much money will each team receive?

Number model: _____

Answer: _____

SRB
22 23

4. Multiply with a paper-and-pencil algorithm.

$66 * 62 = $ _____

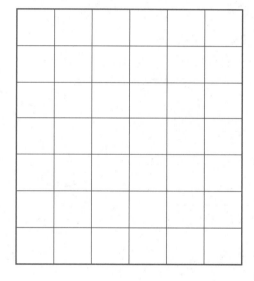

SRB
18 19

5. Complete.

   a. 9 m = _____ cm

   b. 1,500 cm = _____ m

   c. 350 cm = _____ m

   d. 458 cm = _____ m _____ cm

   e. 3.2 m = _____ cm

SRB
129

6. **a.** Shade $\frac{1}{2}$ of the square.

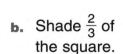

   b. Shade $\frac{2}{3}$ of the square.

SRB
44

LESSON 6·5 Making a Full-Circle Protractor

There are 360 marks around the circle. They divide the edge of the circle into 360 small spaces. Twelve of the marks are longer than the rest. They are in the same positions as the 12 numbers around a clock face. Your teacher will tell you how to label the 12 large marks on the circle.

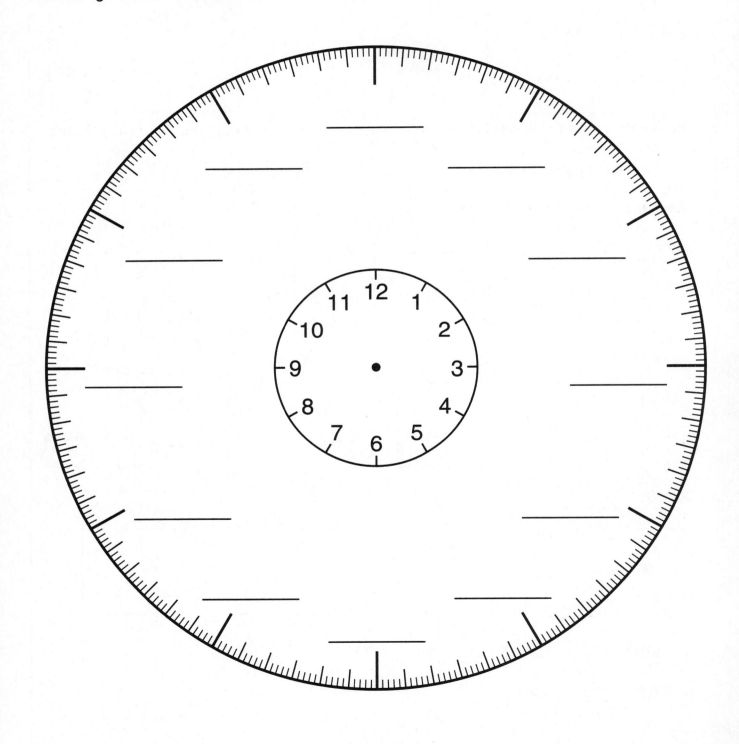

LESSON 6·5 Clock Angles

Use the clock below and the full-circle protractor on journal page 152 to help you answer the questions.

1. How many minutes and how many degrees does the *minute hand* move

   a. from 3:00 to 4:00?   _____ minutes   _____ °

   b. from 7:00 to 7:45?   _____ minutes   _____ °

   c. from 8:15 to 8:45?   _____ minutes   _____ °

   d. from 6:30 to 6:50?   _____ minutes   _____ °

   e. from 5:15 to 5:30?   _____ minutes   _____ °

   f. from 1:00 to 1:10?   _____ minutes   _____ °

   g. from 12:00 to 12:05?   _____ minutes   _____ °

   h. from 5:00 to 5:01?   _____ minutes   _____ °

Try This

2. How many degrees does the *hour hand* move

   a. in 1 hour?   _____ °

   b. in $\frac{1}{2}$ hour?   _____ °

   c. in 10 minutes?   _____ °

3. Explain how you solved Problem 2c.

   _____

   _____

   _____

   _____

LESSON 6·5 Population Bar Graph

The table below shows the percent of the population (number of people out of 100) who are 14 years old or younger in the Region 2 countries.

SRB
76 301

Country	Percent of Population Ages 0–14
France	19
Greece	15
Hungary	16
Iceland	23
Italy	14
Netherlands	18
Norway	20
Poland	18
Spain	15
United Kingdom	19

1. Make a bar graph to display the information given in the table above.

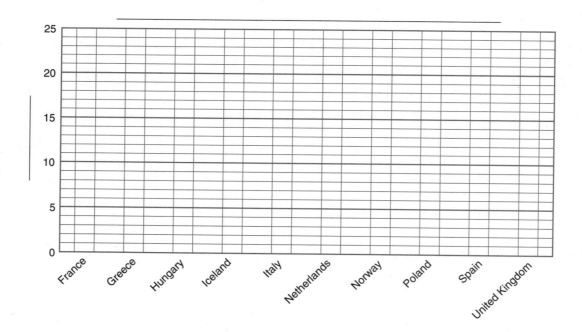

2. Why might it be important to know what percent of the population of a country is 0 through 14 years of age?

_____

LESSON 6·6   **Measuring Angles**

Use your full-circle protractor to measure each angle.

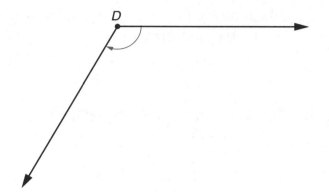

1. ∠C measures _____ °.

2. ∠D measures _____ °.

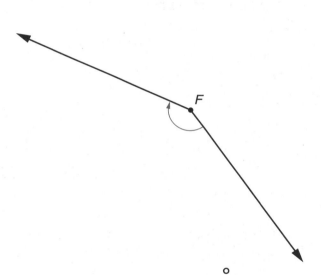

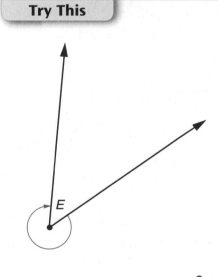

3. ∠F measures _____ °.

4. ∠E measures _____ °.

5. Without using your full-circle protractor, give the measure of the reflex angle in Problem 3 (the part not marked by the blue arrow). Explain your answer.

_____

_____

_____

LESSON 6·6 **Math Boxes**

---

1. Ms. Kawasaki's fourth grade class made a circle graph to show students' favorite days of the week.

   a. Which day of the week is the least favorite in Ms. Kawasaki's classroom?

   _____

   b. About what fraction of the students prefer Saturday?

   _____

Favorite Day of the Week

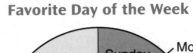

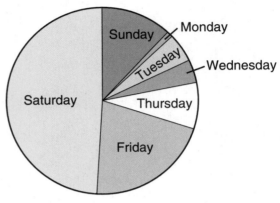

---

2. Juan talked on the phone an average of 34 minutes per week for 1 whole year. About how many minutes did Juan spend on the phone in 1 year?

   Number model: _____

   Answer: _____ minutes

SRB
18 19

3. Divide with a paper-and-pencil algorithm. Write the remainder as a fraction.

   883 / 7 = _____

SRB
22 23
179

---

4. Write <, >, or = to make each number sentence true.

   a. 420,000,000 _____ four hundred twenty million

   b. 65,000,000 _____ 92,000,000

   c. four hundred thousand _____ 10^4

   d. 10^2 _____ 1,000

SRB
5 6

5. For this spinner, what color would you be *most likely* to land on?

   _____

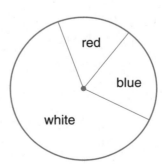

SRB
80 84

LESSON 6·7 Drawing and Measuring Angles

Math Message

Use a straightedge to draw the following angles. Do **not** use a protractor.

∠A: any angle
less than 90°

∠B: any angle more than
90° and less than 180°

∠C: any angle
more than 180°

∠A is called an **acute angle.**   ∠B is called an **obtuse angle.**   ∠C is called a **reflex angle.**

Measuring Angles with a Protractor

Write whether the angle is *acute* or *obtuse*. Then measure it as accurately as you can.

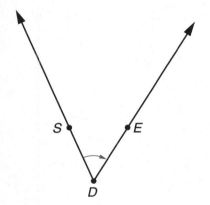

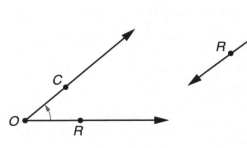

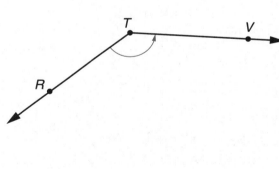

∠SDE is _____.

∠COR is _____.

∠RTV is _____.

∠SDE is about _____ °.

∠COR is about _____ °.

∠RTV is about _____ °.

LESSON 6·7 **Drawing Angles**

1. Draw a 35° angle, using line segment *GH* as one of its sides.

G   H

2. Draw a 150° angle, using ray *CD* as one of its sides.

D   C

3. Draw a 60° angle, using ray *EF* as one of its sides.

F   E

4. Draw a 15° angle, using ray *AB* as one of its sides.

A   B

Try This

5. Draw a 330° angle, using ray *IJ* as one of its sides.

I   J

LESSON 6·7 Math Boxes

1. Insert parentheses to make each number sentence true.

   a. $12 = 15 - 2 + 1$

   b. $66 - 16 * 4 = 200$

   c. $49 = 4 + 3 * 42 / 6$

SRB 150

2. Draw a line segment that is 2 inches long. Mark and label the following inch measurements on the line segment:

$$\frac{1}{4}, \frac{3}{4}, 1, 1\frac{1}{4} \text{ and } 1\frac{1}{2}$$

SRB 128

3. Six classrooms collected newspapers for one week. If they collected a total of 582 newspapers by the end of the week, on average about how many newspapers did each class collect?

Number model: _____

Answer: _____ newspapers

SRB 22 23

4. Multiply with a paper-and-pencil algorithm.

$67 * 34 =$ _____

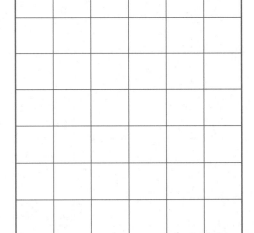

SRB 18 19

5. How many centimeters are in 9.7 meters? Circle the best answer.

   A. 907

   B. 900.7

   C. 970

   D. 9,700

SRB 129 315

6. Circle the square that has $\frac{1}{3}$ shaded.

   A.   **B.**

SRB 44

LESSON 6·8

Math Boxes

1. Name the ordered number pair for each point plotted on the coordinate grid.

A (_____,_____)

B (_____,_____)

C (_____,_____)

D (_____,_____)

E (_____,_____)

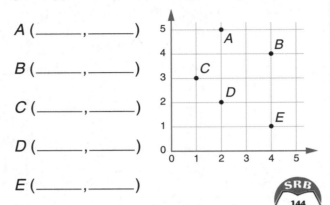

SRB
144

2. Complete the "What's My Rule?" table and state the rule.

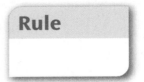

Rule

in	out
3.6	2.1
10	8.5
7.2	
	4.9

SRB
162–166

3. ∠EDF is _____ (acute or obtuse).

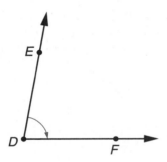

Measure of ∠EDF = _____°.

SRB
93
142 143

4. Cross off the names that do not belong in the name-collection box below.

32
81 − 49
9 * 4
(5 * 6) + 2
98 ÷ 3
10 + 15 + 7

SRB
149

5. Round to the nearest hundred-thousand.

a. 9,540,234 _____

b. 37,609,034 _____

c. 78,291,554 _____

d. 290,696,332 _____

SRB
182 183

6. Fill in the missing fractions on the number line.

_____ _____ _____

SRB
316

LESSON 6·8 A Map of the Island of Ireland

Bantry	B-1	Dublin	F-4	Lahinch	B-4	Omagh	E-7
Belfast	F-7	Dundalk	F-6	Larne	F-7	Tralee	B-2
Carlow	E-3	Galway	C-4	Limerick	C-3	Tuam	C-5
Castlebar	B-6	Gort	C-4	Mullingar	E-5	Westport	B-5
Derry	E-8	Kilkee	B-3	Navan	E-5	Wicklow	F-4

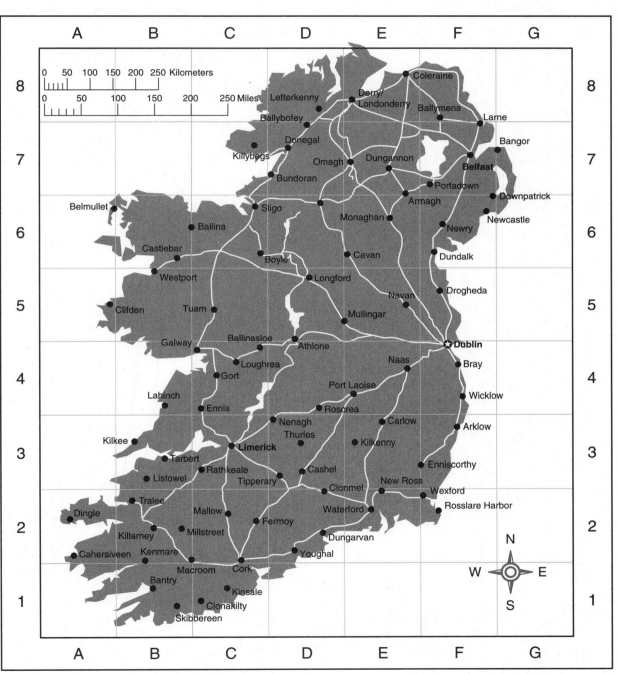

LESSON 6·8

A Campground Map

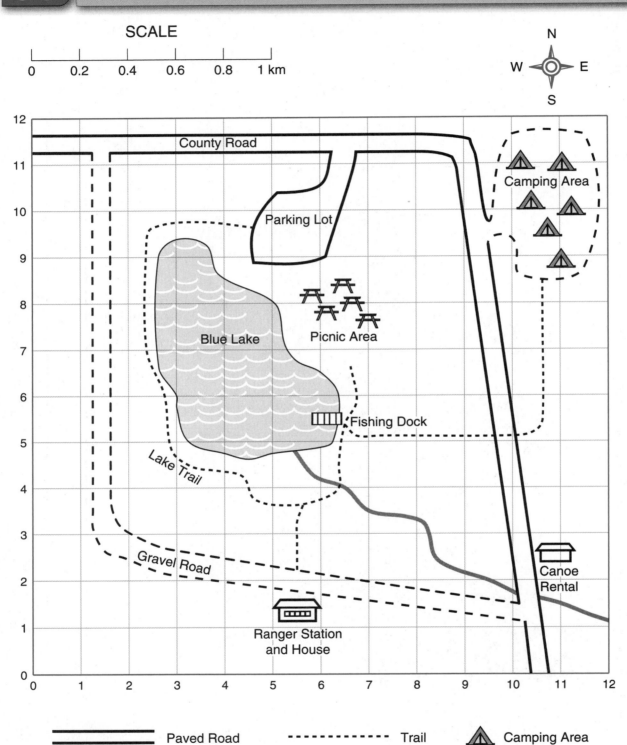

SCALE

| 0 | 0.2 | 0.4 | 0.6 | 0.8 | 1 km |

N
W ← → E
S

Camping Area

Parking Lot

County Road

Blue Lake

Picnic Area

Fishing Dock

Lake Trail

Gravel Road

Canoe Rental

Ranger Station and House

| ═══════ | Paved Road | - - - - - - - | Trail | 🏕 | Camping Area |
| - - - - - | Unpaved Road | ━━━━━ | River | ⛩ | Picnic Area |

LESSON 6·8   Finding Distances on a Map

Use the campground map on journal page 162 to complete the following:

1. Suppose you hiked along the lake trail from the
    fishing dock to the parking lot. Estimate the distance
    you hiked.   About _____ km

2. The ranger made her hourly check. She started at the
    ranger station. She drove northwest and then north
    on Gravel Road to County Road. She turned east
    onto County Road and drove past the parking lot
    and the camping area. After she passed the canoe
    rental, she turned right onto Gravel Road and drove
    back to the ranger station. About what distance did
    she drive?   About _____ km

3. Estimate the distance around Blue Lake.   About _____ km

4. You are planning to hike from the camping area to
    the parking lot. You will stay on the roads or trails.
    You want to hike at least 5 kilometers.

    a. Plan your route. Then draw it on the map with a
       colored pencil or crayon.

    b. Estimate the distance.   About _____ km

5. Use the ordered number pairs to locate each item on the map. Mark a dot
    at the location. Next to the dot, write the letter given for the feature.

Campground Features Chart		
	Location	Letter
parked car	(5,9)	C
boat	$(3\frac{1}{2},8)$	B
swing set	(8,11)	S
hikers	(10.5,6.5)	H
farmhouse	$(\frac{1}{2},7)$	F

LESSON 6·9 — Locating Places on Regional Maps

Use the maps on pages 282–293 in the *Student Reference Book* to answer Problems 1–3.

1. Record the continent in which each city is located.

   a. Pretoria, South Africa (Region 1)   _____

   b. London, England (Region 2)   _____

   c. La Paz, Bolivia (Region 3)   _____

   d. Dhaka, Bangladesh (Region 4)   _____

   e. Washington, D.C., USA (Region 5)   _____

2. Find the approximate latitude and longitude of each city. Record the degrees and circle the correct direction.

   a. Pretoria, South Africa   latitude _____ °N or °S; longitude _____ °E or °W

   b. London, England   latitude _____ °N or °S; longitude _____ °E or °W

   c. La Paz, Bolivia   latitude _____ °N or °S; longitude _____ °E or °W

   d. Dhaka, Bangladesh   latitude _____ °N or °S; longitude _____ °E or °W

   e. Washington, D.C., USA   latitude _____ °N or °S; longitude _____ °E or °W

3. Each degree of latitude that you travel north or south from the equator is equal to about 70 miles. About how many miles from the equator is each city?

   a. Pretoria, South Africa   About _____ miles

   b. London, England   About _____ miles

   c. La Paz, Bolivia   About _____ miles

   d. Dhaka, Bangladesh   About _____ miles

   e. Washington, D.C., USA   About _____ miles

LESSON 6·9   **Math Boxes**

---

1. Cindy received $40 from her aunt and uncle. She drew a circle graph to show how she will use the money.

   a. How much will she save?

   _____

   b. How much will be spent on clothes?

   _____

   c. On movies?

   _____

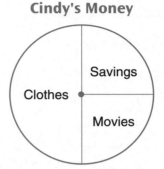

Cindy's Money

Savings

Clothes

Movies

---

2. Mrs. Moy's students are folding paper cranes for an art project. Each of her 27 students is assigned to make at least 15 paper cranes. What is the least number of cranes the class will have for the project?

   Number model: _____

   Answer: _____ paper cranes

3. Divide with a paper-and-pencil algorithm. Write the remainder as a fraction.

   598 / 3 = _____

---

4. Which number sentence is true? Circle the best answer.

   A. $33,000,000 < 33,000$

   B. $5,200,000 > 9$ million

   C. $10^4 = 10,000$

   D. six hundred thousand $= 10^6$

5. For this spinner, which color would you be *least likely* to land on?

   _____

   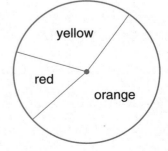

   yellow

   red

   orange

LESSON
6·10   ## Partial-Quotients Division

1. Raul baked 96 cupcakes. He wants to divide them evenly among 3 bake sale tables. How many cupcakes should he put on each table?

   Number model: _____

   Answer: _____ cupcakes

   How many cupcakes will be left over?

   _____ cupcakes

2. The library has boxes to store 132 videotapes. Each box holds 8 tapes. How many boxes will be completely filled?

   Number model: _____

   Answer: _____ boxes

   How many videotapes will be left over?

   _____ videotapes

3. The teacher divided 196 note cards evenly among 14 students. How many note cards did each student get?

   Number model: _____

   Answer: _____ note cards

   How many note cards were left over?

   _____ note cards

4. There are 652 students at a school. The auditorium has rows with 22 seats each. How many rows would be completely filled if all the students attend an assembly?

   Number model: _____

   Answer: _____ rows

   How many students would be left over?

   _____ students

LESSON 6·10

Partial-Quotients Division *continued*

5. 18)864   Answer: _____

6. 509 ÷ 37 = _____

Try This

7. 4,872 / 24 = _____

8. 3,315 ÷ 36 = _____

9. There are _____ players in the league.
(Write a number greater than 300.)

There are _____ players on each team.
(Write a number between 11 and 99.)

How many teams can be made?

Number model: _____

Answer: _____ teams

How many players are left over?

_____ players

LESSON 6·10

Math Boxes

---

1. Name the ordered number pair for each point plotted on the coordinate grid.

A (_____ , _____)

B (_____ , _____)

C (_____ , _____)

D (_____ , _____)

E (_____ , _____)

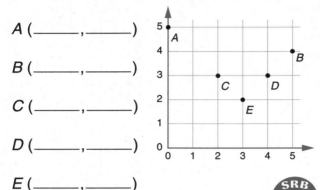

SRB 144

---

2. Complete the "What's My Rule?" table and state the rule.

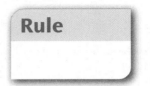
Rule

in	out
3.66	7.04
0.42	3.80
8.73	
	12.66

SRB 162–166

---

3. ∠NMO is _____ (acute or obtuse).

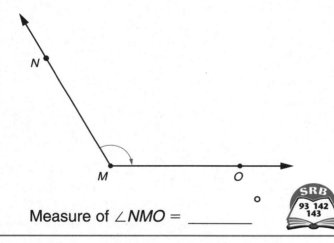

Measure of ∠NMO = _____

SRB 93 142 143

---

4. Cross out the names that do not belong in the name-collection box below.

48
(2 ∗ 3) ∗ 8
100 − 62
18 + 13 + 17
12 ∗ 4
184 ÷ 4

SRB 149

---

5. Round 451,062 to the nearest thousand. Circle the best answer.

A. 500,000

B. 451,000

C. 451,100

D. 452,000

SRB 182 183

---

6. Fill in the missing fractions on the number line.

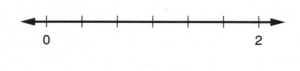

0   2

___ ___ ___ ___ ___

SRB 316

Date _____ Time _____

Math Boxes

1. Fill in the missing fractions on the number lines.

a.

$1\frac{2}{5}$ $2\frac{2}{5}$

_____   _____   _____   _____

b.

$2\frac{1}{2}$ 5

_____   _____   _____   _____

2. Draw 12 balloons. Circle $\frac{5}{12}$ of the balloons. Mark X on $\frac{1}{4}$ of the balloons.

SRB
44

3. Write five names for $\frac{1}{4}$.

SRB
149

4. a. Shade $\frac{5}{6}$ of the hexagon.

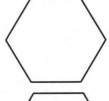

b. Shade $\frac{2}{3}$ of the hexagon.

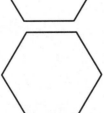

SRB
44

5. Design a spinner such that it is more likely that you will land on red than on green.

SRB
84

169

My Route Log

Date	Country	Capital	Air distance from last capital	Total distance traveled so far
	1 U.S.A.	Washington, D.C.		
	2 Egypt	Cairo		
	3			
	4			
	5			
	6			
	7			
	8			
	9			
	10			
	11			
	12			
	13			
	14			
	15			
	16			
	17			
	18			
	19			
	20			

Route Map

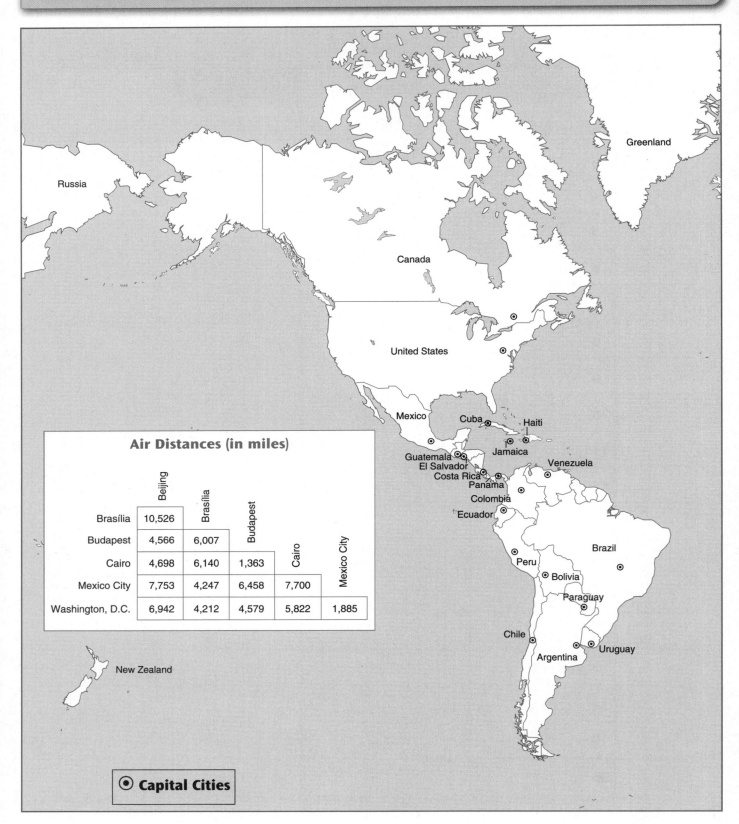

Air Distances (in miles)

	Beijing	Brasília	Budapest	Cairo	Mexico City
Brasília	10,526				
Budapest	4,566	6,007			
Cairo	4,698	6,140	1,363		
Mexico City	7,753	4,247	6,458	7,700	
Washington, D.C.	6,942	4,212	4,579	5,822	1,885

⊙ **Capital Cities**

Russia

Greenland

Canada

United States

Mexico

Cuba

Haiti

Guatemala

Jamaica

El Salvador

Venezuela

Costa Rica

Panama

Colombia

Ecuador

Brazil

Peru

Bolivia

Paraguay

Chile

Uruguay

Argentina

New Zealand

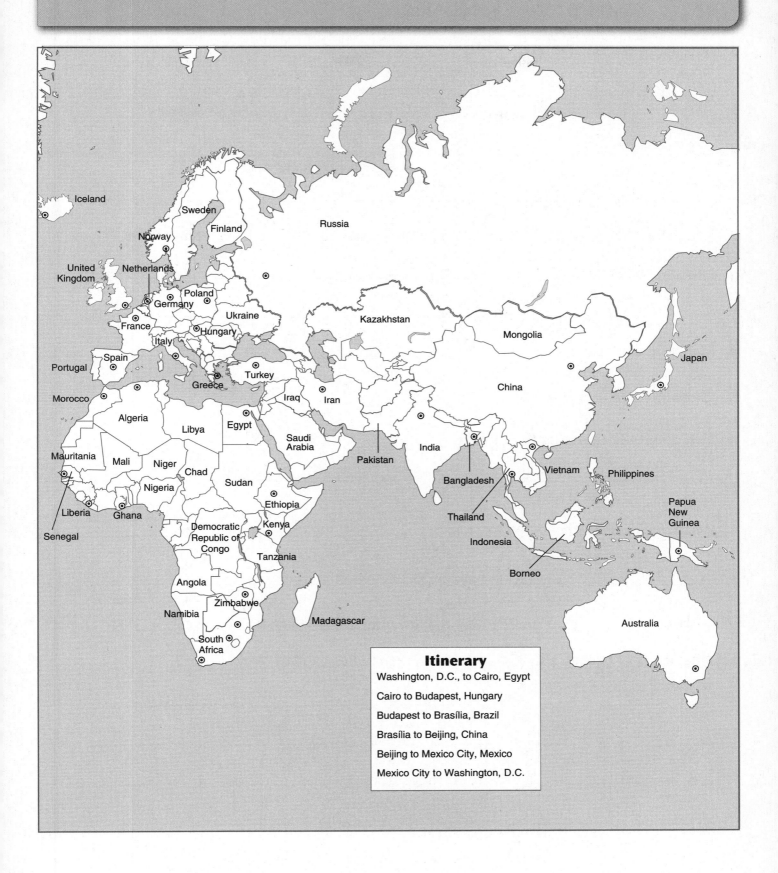

Itinerary

Washington, D.C., to Cairo, Egypt

Cairo to Budapest, Hungary

Budapest to Brasília, Brazil

Brasília to Beijing, China

Beijing to Mexico City, Mexico

Mexico City to Washington, D.C.

LESSON 3·6 My Country Notes

A. **Facts about the country**

_____ is located in _____.
          name of country   name of continent

1. It is bordered by _____
                                          countries, bodies of water

_____.

2. Population: _____   Area: _____ square miles

3. Languages spoken: _____

_____

4. Monetary unit: _____

5. Exchange rate (optional): 1 _____ = _____

B. **Facts about the capital of the country**

_____ Population: _____
          name of capital

1. When it is noon in my hometown, it is _____ in _____.
                                                          time (A.M. or P.M.?)   name of capital

2. In _____/_____, the average high temperature in _____
        month   month   name of capital

is about _____°F. The average low temperature is about _____°F.

3. What kinds of clothes should I pack for my visit to this capital? Why?

_____

_____

_____

_____

LESSON 3·6 **My Country Notes** *continued*

4. Turn to the Route Map found on journal pages 172 and 173.
   Draw a line from the last city you visited to the capital of this country.

5. If your class is using the Route Log, record the information on journal page 171 or
   Math Masters, page 421.

6. Can you find any facts on pages 302–305 in your *Student Reference Book* that
   apply to this country? For example, is one of the 10 tallest mountains in the world
   located in this country? List all the facts you can find.

   _____

   _____

   _____

   _____

c. My impressions about the country

   Do you know anyone who has visited or lived in this country? If so, ask that person
   for an interview. Read about the country's customs and about interesting places to
   visit there. Use encyclopedias, travel books, the travel section of a newspaper, or
   library books. Try to get brochures from a travel agent. Then describe below some
   interesting things you have learned about this country.

   _____

   _____

   _____

   _____

   _____

LESSON 4·7   My Country Notes

A. Facts about the country

_____ is located in _____ .
name of country   name of continent

1. It is bordered by _____

   countries, bodies of water

   _____ .

2. Population: _____   Area: _____ square miles

3. Languages spoken: _____

   _____

4. Monetary unit: _____

5. Exchange rate (optional): 1 _____ = _____

B. Facts about the capital of the country

_____   Population: _____
name of capital

1. When it is noon in my hometown, it is _____ in _____ .

   time (A.M. or P.M.?)   name of capital

2. In _____/_____, the average high temperature in _____

   month   month   name of capital

   is about _____°F. The average low temperature is about _____°F.

3. What kinds of clothes should I pack for my visit to this capital? Why?

   _____

   _____

   _____

   _____

176

LESSON 4·7

My Country Notes *continued*

4. Turn to the Route Map found on journal pages 172 and 173.
   Draw a line from the last city you visited to the capital of this country.

5. If your class is using the Route Log, record the information on journal page 171 or *Math Masters,* page 421.

6. Can you find any facts on pages 302–305 in your *Student Reference Book* that apply to this country? For example, is one of the 10 tallest mountains in the world located in this country? List all the facts you can find.

_____

_____

_____

_____

_____

c. My impressions about the country

Do you know anyone who has visited or lived in this country? If so, ask that person for an interview. Read about the country's customs and about interesting places to visit there. Use encyclopedias, travel books, the travel section of a newspaper, or library books. Try to get brochures from a travel agent. Then describe below some interesting things you have learned about this country.

_____

_____

_____

_____

_____

_____

_____

LESSON 5·11 My Country Notes

A. Facts about the country

_____ is located in _____.
            name of country   name of continent

1. It is bordered by _____
                                    countries, bodies of water

   _____.

2. Population: _____   Area: _____ square miles

3. Languages spoken: _____

   _____

4. Monetary unit: _____

5. Exchange rate (optional): 1 _____ = _____

B. Facts about the capital of the country

_____   Population: _____
            name of capital

1. When it is noon in my hometown, it is _____ in _____.
                                                time (A.M. or P.M.?)   name of capital

2. In _____/_____, the average high temperature in _____
          month   month   name of capital

   is about _____°F. The average low temperature is about _____°F.

3. What kinds of clothes should I pack for my visit to this capital? Why?

   _____

   _____

   _____

   _____

LESSON 5·11 My Country Notes *continued*

4. Turn to the Route Map found on journal pages 172 and 173.
Draw a line from the last city you visited to the capital of this country.

5. If your class is using the Route Log, record the information on journal page 171 or
Math Masters, page 421.

6. Can you find any facts on pages 302–305 in your *Student Reference Book* that
apply to this country? For example, is one of the 10 tallest mountains in the world
located in this country? List all the facts you can find.

_____

_____

_____

_____

_____

C. My impressions about the country

Do you know anyone who has visited or lived in this country? If so, ask that person
for an interview. Read about the country's customs and about interesting places to
visit there. Use encyclopedias, travel books, the travel section of a newspaper, or
library books. Try to get brochures from a travel agent. Then describe below some
interesting things you have learned about this country.

_____

_____

_____

_____

_____

_____

_____

LESSON 6·7 | **My Country Notes**

A. Facts about the country

_____ is located in _____.

name of country   name of continent

1. It is bordered by _____

countries, bodies of water

_____.

2. Population: _____ Area: _____ square miles

3. Languages spoken: _____

_____

4. Monetary unit: _____

5. Exchange rate (optional): 1 _____ = _____

B. Facts about the capital of the country

_____ Population: _____

name of capital

1. When it is noon in my hometown, it is _____ in _____.

                                      time (A.M. or P.M.?)   name of capital

2. In _____/_____, the average high temperature in _____

    month   month   name of capital

is about _____°F. The average low temperature is about _____°F.

3. What kinds of clothes should I pack for my visit to this capital? Why?

_____

_____

_____

_____

LESSON 6·7   My Country Notes *continued*

4. Turn to the Route Map found on journal pages 172 and 173.
   Draw a line from the last city you visited to the capital of this country.

5. If your class is using the Route Log, record the information on journal page 171 or
   Math Masters, page 421.

6. Can you find any facts on pages 302–305 in your *Student Reference Book* that
   apply to this country? For example, is one of the 10 tallest mountains in the world
   located in this country? List all the facts you can find.

_____

_____

_____

_____

_____

C. My impressions about the country

Do you know anyone who has visited or lived in this country? If so, ask that person
for an interview. Read about the country's customs and about interesting places to
visit there. Use encyclopedias, travel books, the travel section of a newspaper, or
library books. Try to get brochures from a travel agent. Then describe below some
interesting things you have learned about this country.

_____

_____

_____

_____

_____

_____

_____

*,/ Fact Triangles 1

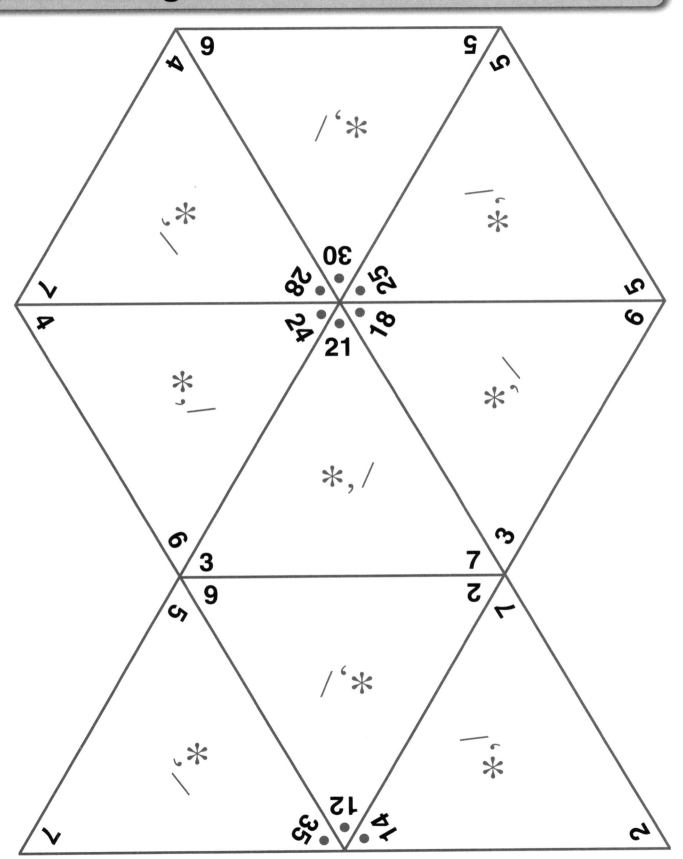

*,/ **Fact Triangles 2**

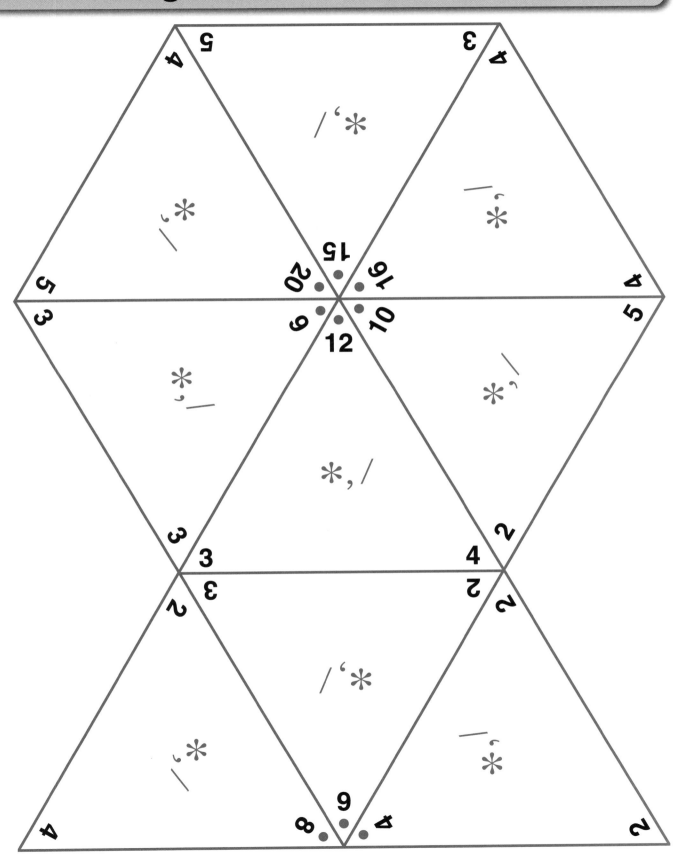

*,/ **Fact Triangles 3**

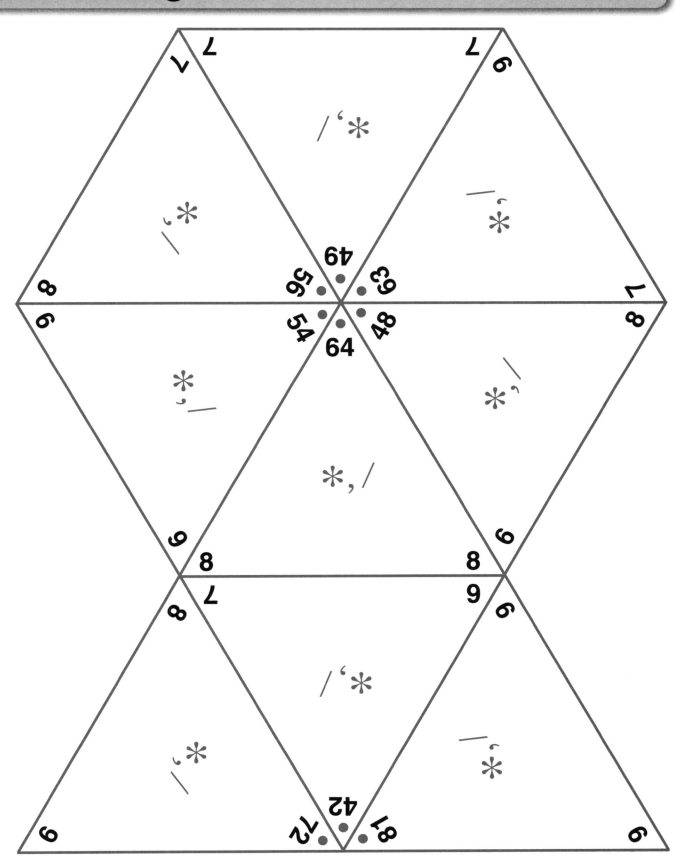

*,/ **Fact Triangles 4**

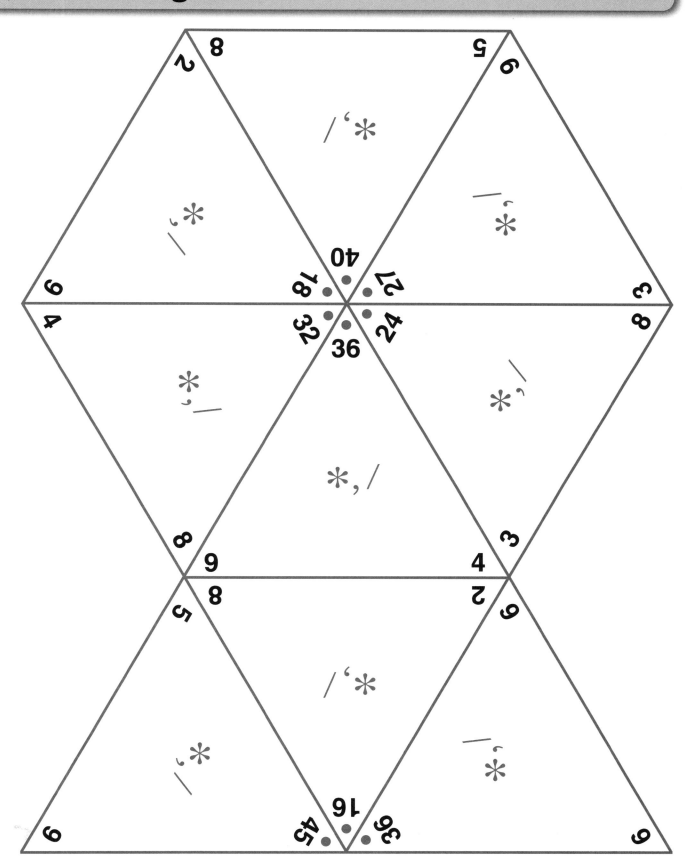